Artificial Intelligence and the Future of Humanity

Lab Maharaj

DEDICATION

This book is dedicated to the one I love, who makes every day an adventure. Thank you for being my everything, I couldn't ask for anything more.

ACKNOWLEDGMENTS

Writing a book is an incredibly difficult and rewarding experience. It is only fitting that I acknowledge those who have made this journey possible. Without the guidance, support, and encouragement of my family, friends, mentors, and peers, I wouldn't be here today. They have all been essential to me in achieving this dream of being a published author.

Introduction

The rise of the machines is upon us. Artificial Intelligence (AI) has been steadily making its presence known in our world, and it's changing the way we work, think and live. AI technology has grown rapidly over recent years as its capacity to understand human behaviour, problem-solve complex tasks and interact with people grows exponentially. It has already infiltrated many aspects of everyday life - from healthcare to transportation - leaving no industry untouched.

AI can be used for a variety of tasks; from automating mundane processes to streamlining complex operations. By relying on this technology, a range of benefits can be achieved; such as increased efficiency and accuracy, reduced costs associated with production or services and better utilisation of resources. While concerns exist around potential job losses due to automation, new roles are also being created that require specialised skill sets in order to manage and optimise AI systems.

As AI becomes more advanced and ubiquitous, it is increasingly being used for a wide range of applications across industries such as healthcare, finance, transportation and security. Machine learning algorithms are becoming increasingly sophisticated as they are trained on larger datasets than ever before. This means that machines can now solve problems humans once thought were impossible to solve using traditional methods. As a result, AI technologies are changing how businesses operate and how people interact with machines on a daily basis.

This book offers readers an introduction to AI technologies such as machine learning, natural language processing and deep learning. Written by a specialist in the field, it covers topics such as designing intelligent systems; understanding data-driven models; building autonomous agents; and exploring ethical implications surrounding this technology.

Historical Development

The "Rise of the Machines" is a phenomenon which has been seen in recent decades as technological advances have continued to proliferate, resulting in increased automation and artificial intelligence. This trend has had a significant impact on labour markets, with the displacement of human workers in favour of machines, thus reducing labour costs and increasing efficiency. In addition, this shift toward automated technologies has enabled organisations to reach new heights of productivity and product quality, due to the advanced capabilities of these technological systems.

The Historical Development of AI has been steadily progressing since the 1950s. Artificial Intelligence (AI) is defined as the ability of a computer system to think and learn in a way that mimics human intelligence. Over the past 70 years, AI technologies have advanced significantly, leading to applications in several industries such as healthcare and automotive.

In 1956, John McCarthy first used the term "Artificial Intelligence" at a research conference at Dartmouth College. This event is widely credited with kick-starting an entire field of study devoted to developing machines that could replicate some aspects of human behaviour. Subsequently, significant advancements continued throughout the decades by researchers from all over the world who sought to build upon existing technology and create something even more innovative.

Early Beginnings: Turing's Test

The world of Artificial Intelligence (AI) has experienced astronomical growth in recent years. With the invention of modern machine learning, AI has become a tool that can provide solutions to complex problems. This amazing technology is the result of decades of research and development by many brilliant minds, both visionaries and scientists alike.

In particular, pioneers like Alan Turing, Marvin Minsky and John McCarthy have contributed greatly to early advancements in AI technology. Turing is widely regarded for proposing the 'Turing Test' which remains an integral part of the Turing Machine Theory developed in 1936. Similarly, Minsky and McCarthy are credited with introducing some pioneering concepts such as neural networks and Lisp programming language respectively in the 1950s-1960s.

The Turing Test, first proposed by Alan Turing in 1950, is a landmark moment in the history of artificial intelligence. It sought to answer the fundamental question: can machines think? The test calls for a human judge to interact with two agents - one is a machine and the other is another human. If the judge cannot tell which agent is which, then the machine passes the test.

Turing's original paper on his proposal was revolutionary at its time and sparked off many debates about what makes us uniquely human. His ideas on how machines could be used to simulate intelligent behaviour have been adapted and improved upon over time, leading to modern advances in AI technology such as natural language processing and autonomous robots.

His 1936 paper, "On Computable Numbers with an Application to the Entscheidungsproblem," laid out his ideas for how a machine could use instructions to manipulate symbols. This laid the groundwork for modern computer science and paved the way for advancements in artificial intelligence.

Turing's paper also developed what is now known as the "Turing Test", which is a benchmark for determining whether or not a machine can exhibit behaviours that are indistinguishable from those of a human being. This test has been used to measure AI capabilities ever since its inception, and continues to be highly relevant today as technology advances further into AI research.

The test was designed to assess the capabilities of a computer's ability to exhibit intelligent behaviour equivalent to that of a human. The test consists of an interrogator who interacts with two subjects, one a human and one a computer, without being able to see either subject. This interaction is meant to determine whether the computer can successfully imitate and demonstrate human conversation.

1950s: Birth of AI

The 1950s marked a crucial period in the history of Artificial Intelligence (AI). In 1950, Alan Turing published his paper 'Computing Machinery and Intelligence', which was a milestone in the birth of AI. It set forth the Turing Test as a way to determine whether machines are capable of thinking similar to humans. This led to further research into AI and its potential applications.

In 1956, John McCarthy coined the term Artificial Intelligence at the Dartmouth Summer Research Project on Artificial Intelligence. This seminal event kickstarted the study of AI and ushered in an era that saw tremendous progress in this field. The late 50s saw many advances such as pattern recognition systems, self-correcting programs, natural language processing capabilities and robots with some degree of autonomy.

Pattern Recognition Systems

Pattern recognition systems have played an integral role in the history of artificial intelligence. It is widely accepted that the birth of AI began with work done to develop pattern recognition algorithms and machines over 60 years ago. Early experiments with neural networks demonstrated a capacity for recognizing patterns that had never before been seen. Through these early successes, researchers developed a passion for finding ways to make computers "smarter" and more capable of performing complex tasks quickly and accurately.

The development of pattern recognition systems has come leaps and bounds since the early days of AI research. Algorithms are now capable of learning from data sets, processing images and videos quickly, as well as taking on many other difficult tasks such as natural language processing (NLP). Furthermore, recent advances in machine vision have enabled computers to interpret scenes more accurately than ever before.

Self-correcting programs are a type of AI software designed to learn from its mistakes and improve itself over time without requiring any intervention by a human programmer.

These types of programs have been used for decades in many areas such as robotics, security systems, and financial analysis. Self-correcting programs use algorithms that allow them to analyse their performance over time and adjust their behaviour accordingly. This enables the program to adapt quickly to new situations and keep up with changing trends in its environment. By continually learning from its mistakes, the program can become more efficient and accurate with each iteration it makes.

Natural language processing (NLP) capabilities have been a cornerstone of artificial intelligence technology since its inception. NLP is the study of how computers process and understand human language, allowing for more fluid communication between humans and machines. The idea has been around since the 1950s when Alan Turing proposed a mathematical model that could test a machine's ability to think like a human being. This was part of his famous "Turing Test," which spurred research in the field of AI for decades to come. The mathematical model that could test a machine's ability to think like a human being. The model was designed to assess whether or not machines can be deemed "intelligent" and it has become known as the Turing Test.

The test consists of an interrogator who would interact with two entities: one human and one machine. If the interrogator couldn't distinguish between them, then the machine passed the test and could be deemed intelligent. Although this model has been criticised over time since it may not accurately measure intelligence, it did lead to incredible advancements in computer science, paving the way for further research into AI technologies today.

Pioneers such as Frank Rosenblatt, recognized the potential of using computers to analyse data and make decisions. He created the Perceptron, an early form of machine learning capable of identifying objects. This breakthrough represented a major step forward in understanding how computer algorithms could be used to recognize patterns.

Other researchers, such as Marvin Minsky and John McCarthy, contributed to the development of AI by defining its principles and exploring ways for machines to "learn" from their environment. They formulated the idea that computers should be able to replicate human behaviour such as decision-making and problem-solving. Their research spawned an entire field devoted to artificial intelligence which is still growing today with new advancements being made all the time.

Science Fiction: Robots

Robots with some degree of autonomy have been the subject of science fiction for centuries. However, it was not until the mid-twentieth century that these robots began to appear in reality. These computers were to mimic humans and make decisions on their own. By the 1960s, robots began to embody more complex self-regulating systems as they became capable of completing tasks without human intervention and were able to use sensors and processors to observe environments, identify objects, and adjust behaviours accordingly. Automation has been around since ancient times, but it wasn't until the 1940s that robotic sensing technology started to play a role in the development of AI. The invention of computers and advancements in technology enabled scientists to finally explore AI as an area of research.

The first breakthrough in robotic sensing came with the invention of optical sensors. These sensors allowed robots to detect objects and respond accordingly, making them much more intelligent than before. This was followed by further developments such as tactile sensors, which allowed robots to feel objects and react appropriately; proximity sensors, which enabled robots to detect nearby obstacles; and motion detectors, which let bots move around without human intervention.

70s-80s: AI Boom & Bust

The 1970s and 1980s marked a period of growth and development in the field of Artificial Intelligence (AI). In the early years, researchers were filled with optimism as they believed that AI could bring about revolutionary advancements in technology. However, by the mid-1980s, this optimism had begun to wane amidst public scepticism towards AI.

At the start of the decade, computer scientists sought to build machines capable of performing tasks previously deemed impossible. This goal quickly became attainable due to significant breakthroughs made in artificial intelligence research. Companies such as IBM began investing heavily in AI projects such as natural language processing and robotics. As a result, this boom led to an immense amount of progress within the field during those two decades.

Unfortunately, these advancements failed to live up to their lofty expectations leading many investors and companies to abandon AI projects entirely by 1985.

The groundwork for modern AI began with the invention of the first digital computer program by Alan Turing during World War II. This set off a wave of research into the field, with scientists exploring new ways to create more efficient machines capable of self-learning. These advancements eventually led to breakthroughs such as neural networks, which allow computers to learn from data without any human input, and natural language processing, which enables machines to understand human language.

By the late 70s and early 80s, AI had advanced significantly, allowing it to be applied practically in many industries such as finance and healthcare.

AI was primarily used to automate mundane tasks like data processing and routine decision making. Companies soon realised they could use AI to improve their operations by introducing algorithms into their systems that could analyse customer trends or create predictive models. With advancements in technology, AI is now capable of carrying out more complicated tasks like image recognition or natural language processing. These days we can see its application in autonomous vehicles, voice assistants, facial recognition software, and other everyday applications.

AI could be used to identify patterns in customer behaviour, such as buying habits or usage patterns. Companies could then use this data to develop strategies that better served their customers' needs. They could target specific segments of the market with tailored offerings and create more effective marketing campaigns. This enabled them to determine when certain products should be offered and at what price points. By using AI for customer trend analysis, companies were able to maximise their profits by understanding their customers better.

The early days of AI during this period were focused on robotics and natural language processing. Researchers strove to create machines that could understand and respond to human speech, as well as move autonomously with some degree of dexterity. One such robot was Shakey the Robot – developed by Stanford Research Institute for the American Department of Defense – that was capable of navigating through an environment without any external guidance or control.

Developed by Stanford Research Institute (SRI), Shakey was able to move around its environment autonomously while also responding to commands given verbally or through a keyboard.

Shakey's design was based on an integrated system that combined various components such as: perception, reasoning, learning and decision-making abilities. It employed various state-of-the-art technologies such as natural language processing capabilities for understanding vocal commands and image recognition for interpreting visual information. Additionally, it had several actuators allowing it to move across surfaces with ease.

90s-2000s: Reemergence & Research

The 90s-2000s era saw a revolutionary reemergence of artificial intelligence research. In the decades that followed, technological advancements were made, allowing artificial intelligence to reach new heights. During this pivotal time in AI history, scientists and engineers worked fervently to create algorithms and systems that could outperform humans in fast-paced decision making processes.

This period was marked by headway advancements for AI technology, such as machine learning, deep learning networks and natural language processing tools. These systems allowed computers to learn from data without the need for programming techniques. With the help of these technologies, machines began taking on more complicated tasks with relatively greater accuracy than humans ever had before.

Today, artificial intelligence continues to be an integral part of modern society – from medical diagnosis applications to self-driving vehicles – the possibilities are seemingly endless!

By the turn of the millennium, AI had become an area of intense scrutiny. For example, experts sought out ways that AI could be used to automate mundane tasks and make decision-making processes faster and more efficient. Beyond this practical application, researchers also explored what could be done with machine learning algorithms; they examined how computers might be able to interpret data in increasingly complex ways. In addition, many scholars probed deeper questions about computer consciousness and moral autonomy.

There was a focus on developing algorithms that could learn from experience and make decisions based on real-world data. This development was driven by advances in computing technologies and techniques such as neural networks, machine learning, natural language processing, computer vision and robotics. By leveraging these new AI capabilities, researchers were able to develop more effective models for decision-making and problem solving.

In addition to research into algorithms for learning from experience, there was also an increased focus on understanding how humans process information. With this knowledge, scientists were able to create cognitive architectures which incorporated elements of human thought processes into AI systems. These architectures allowed for more sophisticated solutions than traditional programming methods allowed for previously.

This work has seen a rapid expansion due to advancements in computer science and neuroscience. As computers become better at learning tasks that were previously thought impossible for them to do, researchers are using this technology to gain insight into what makes us human. The resulting data provides valuable information about the inner workings of our cognitive processes and neural networks. This understanding can then inform the development of more effective machine-learning algorithms that enable machines to function more like humans do.

Humans have long been interested in the relationship between cognition and Artificial Intelligence (AI). From early attempts to create computers that could think like humans to modern advances in machine learning, scientists have sought to bridge the gap between cognitive processing and AI.

Cognitive functions of humans involve a range of abilities that are essential for day-to-day life. These include memory, attention, problem solving, decision making, language understanding and production, motor control and coordination. Historically, attempts have been made to replicate these cognitive processes with AI systems. With advancements in AI technology such as deep learning algorithms and natural language processing (NLP) methods over the past decade or so, various aspects of cognitive functioning can now be simulated more accurately by computer systems than ever before.

2010s: AI Revolution

The 2010s have been a decade of incredible advances in Artificial Intelligence (AI). From improved algorithms to neural networks, AI has moved from an obscure topic to a major part of everyday life. With applications ranging from facial recognition to self-driving cars, the impact of this technology can be felt in almost every industry.

AI has been around since the 1950s, but only in the last ten years have researchers and developers made significant strides towards creating more sophisticated systems that could operate independently with limited human intervention. From machine learning techniques such as deep learning and reinforcement learning to natural language processing and computer vision, AI has come a long way since its humble beginnings. This revolution has led to breakthroughs in healthcare, finance, retail, transportation and many other fields where automation is now commonplace.

Among the most popular branches of AI are deep learning and reinforcement learning, both of which have had a long and storied history within the field.

Deep learning is a subset of Artificial Intelligence (AI) that has been around since the 1950s. It is an advanced technique used to process complex data, achieve better insights and automate mundane tasks. Deep learning works by using model networks of layers which are trained to recognize patterns and build their own rules from the data they analyse.

Deep learning is a form of machine learning that is based on artificial neural networks; it works by identifying patterns in data sets through successive layers or "deep" layers of analysis. This method is particularly useful for complex tasks such as image recognition and natural language processing. Reinforcement learning, meanwhile, uses rewards to teach machines how to solve problems. This type of AI requires an agent to interact with an environment in order to receive feedback that can be used to update its approach and improve its performance over time.

Natural language processing (NLP) and computer vision are two important fields of artificial intelligence. NLP is the study of how humans use and understand language, while computer vision focuses on enabling computers to interpret and understand visual information from images or videos.

The history of these two technologies dates back to the early 1950s when Alan Turing wrote a paper about artificial intelligence which laid out his famous Turing test for machines to demonstrate their ability to think. Later in the 1960s, NLP began with experiments by linguists at MIT who developed programs that could analyse natural language for grammar and sentence structure. Computer vision first emerged in 1968 when University of Toronto professor Winston Zadeh proposed "fuzzy logic" which allowed computers to process data more like humans do.

Fuzzy logic is an integral part of Artificial Intelligence (AI). It is a form of reasoning that mimics the way humans make decisions, allowing machines to solve complex problems.

Fuzzy logic has been used in AI since its inception in 1965, when Lotfi Zadeh first proposed the concept at the IEEE Conference on Decision and Control.

The goal of fuzzy logic was to bridge the gap between traditional AI and human decision making processes. It does this by creating various membership functions which assign values to data points based on their relative position compared to specific thresholds. For example, if a temperature reading falls outside of a certain range, it may be flagged as either "hot" or "cold" depending on how far away it is from those thresholds.

Artificial Intelligence has come a long way since its first use by Alan Turing in 1950. It has rapidly developed over the years, allowing us to explore and discover even more of what AI can do. Although there is still much to learn and understand about AI, its historic development provides important insights that can inspire further advances. By understanding AI's history and application, we can better understand the impact it will have on our lives in the future.

Economic Benefits

The introduction of artificial intelligence (AI) into the economy has provided numerous benefits to businesses, consumers and governments. AI is a powerful tool that automates processes and tasks across multiple industries, allowing businesses to save time and money while improving the customer experience. Additionally, AI has enabled governments to collect data for economic analysis, resulting in more accurate economic forecasts and better decision-making.

In terms of business operations, AI can streamline processes such as production scheduling or inventory management. By reducing manual labour costs and increasing efficiency levels, companies can lower their overhead expenses significantly. Furthermore, AI technologies are also capable of gathering massive amounts of customer data which can be leveraged for marketing initiatives or product development purposes. This allows companies to better understand consumer needs and provide personalised services that meet those needs.

The use of artificial intelligence (AI) has been increasing rapidly in recent years, and its economic benefits are becoming increasingly clear. AI is a powerful tool for businesses, allowing them to automate mundane tasks, improve their decision-making processes, and increase efficiency. This not only helps businesses save money on labour costs but also boosts productivity and drives innovation.

On the macroeconomic level, AI can help reduce unemployment by creating new jobs in fields that require both technical knowledge and human skills. It can also help increase GDP growth by providing more efficient ways to produce goods and services with fewer resources. Additionally, AI enables companies to better analyse large amounts of data quickly, which can lead to more informed decisions when it comes to policymaking or investments. In this way, AI can be an invaluable asset for governments as they strive towards smarter economic policies.

Positive Effects: Automation & Efficiency

Automation and efficiency are two important concepts in the modern economy. With artificial intelligence (AI) becoming more prevalent, businesses of all sizes can benefit from increased automation and efficiency. AI technology helps to streamline processes, improve customer service, reduce costs, and ultimately drive profits.

By introducing automation into various aspects of the business model, operators can gain technological advantages that increase their production speed and capacity. Efficiency is about optimising existing workflows for maximum performance output with minimal effort. This means utilising the most effective resources available to complete a job faster while maintaining quality levels. With AI technology, companies can automate repetitive tasks and determine which processes are best suited for automation or manual labour depending on the task at hand. This allows organisations to allocate resources more effectively by focusing on higher value-added activities instead of mundane ones that could easily be automated with AI algorithms.

The use of automation and artificial intelligence has become increasingly popular in the business world, leading to more efficient processes that help improve a company's economic bottom line. Automation and artificial intelligence are allowing businesses to speed up production times, reduce labour costs, increase accuracy and consistency in their products, and open the door for new opportunities by allowing them to focus on other aspects of the industry.

Automation is increasing efficiency exponentially. In fact, some experts predict that automation will be able to take over many time-consuming tasks currently done by humans. This can lead to improved overall productivity by eliminating tedious manual labour activities or repetitive jobs that need multiple people working together. Additionally, it also means faster turnarounds on projects with fewer errors due to reduced human error within the process.

Economists have argued that automation can enhance economic growth by creating new opportunities for businesses and workers alike. Automation can help to reduce processing costs in manufacturing, increase efficiency in communication, improve accuracy in data processing, automate repetitive tasks such as customer service inquiries or data entry, create more complex products faster and better than manual labour could ever do before - ultimately leading to higher profits for businesses.

Automation has been gaining traction in macroeconomics and is revolutionising the way economists study economic systems. Artificial intelligence (AI) algorithms are being used to make complex decisions quickly, allowing economists to analyse large amounts of data more accurately than ever before. AI-based models can be trained on historical data and then used to forecast how changing market conditions or policies will impact an economy.

These AI-powered tools have made it possible for economists to test theories much faster than before and gain insights into the functioning of economies with unprecedented precision. This allows them to identify patterns that were once difficult or impossible to detect, leading to new ways of understanding and predicting economic activity. Automation is also improving risk management processes by helping assess the potential effects a policy decision might have on an economy's overall performance.

It is rapidly changing the landscape of microeconomics. With artificial intelligence (AI) at the helm, microeconomists are able to gain a better understanding of how markets work and what factors contribute to their behaviour. AI can be used in various ways to explore the conditions that characterise different types of economies and provide more accurate solutions for economic situations.

AI-driven automation has become an essential tool in microeconomic research and analysis. By analysing large datasets and utilising sophisticated algorithms, AI can identify patterns in historical data and make predictions about future market trends. Furthermore, AI enables economists to make more informed decisions based on complex data sets, making it easier for them to devise strategies that are tailored to a particular market or situation. For example, AI-driven models can help detect pricing anomalies in markets which could indicate potential problems or opportunities.

AI has been quickly integrated into many aspects of modern life, with a wide range of applications across industries. One area in which AI can have meaningful and positive effects is economics – this technology provides powerful tools to assist with economic decision-making and forecasting.

AI-driven systems can provide valuable insights for economists, enabling them to make more informed decisions. By using AI algorithms, economists can conduct simulations that simulate real-world conditions and help determine the likely outcomes of different economic policies or strategies. For example, an AI system could be used to measure the expected impacts of changes in monetary policy on macroeconomic performance such as inflation or GDP growth.

Venture capitalists are investing heavily in artificial intelligence (AI) investments as the technology revolutionises entire industries. By leveraging AI, companies can streamline their operations, increase efficiency, and improve customer experience. This has led to a surge in venture capital investment for AI startups.

The influx of venture capital into the AI market has had a positive impact on economics. With more money available for projects related to artificial intelligence, research efforts have increased exponentially and dozens of new products and services have been developed that promise to revolutionise how businesses operate and how people interact with each other. As a result, these investments are driving economic growth by creating jobs, stimulating demand for services and products, and providing new sources of revenue.

By using AI to analyse data, identify problems and predict outcomes, economists can create more effective economic policies. AI can be used to develop new insights into economic trends, such as stock market movements, identifying potential investments and making forecasts about future economic conditions. AI can also help governments make better decisions about taxation and public spending by providing real-time feedback on how policies impact the economy.

AI's ability to process huge amounts of data quickly makes it a powerful tool for understanding global economics. It can provide detailed analysis of macro-level trends such as consumer spending habits or long-term changes in GDP growth. Economists are able to use this information to create more effective strategies for improving economies over time.

Given the importance of artificial intelligence in guiding economic development and decision-making, researchers have explored its implications for economics. For example, studies show that AI can improve financial forecasting accuracy and reduce transaction costs. It can also help to identify market trends and detect anomalies in stock prices or currency exchange rates. Additionally, AI can be used to explore areas such as corporate governance and public policy development.

The rise of AI has huge potential for economic success, but it comes with a unique set of challenges too. Artificial intelligence can generate large amounts of data which could be misused or cause disruptive market changes if not properly managed.

Quality of Life Improvements

Artificial intelligence has already had a major impact on the economy, and its potential to improve quality of life is only beginning to be explored. AI technology is being applied in many different ways that can increase prosperity across various sectors of society. As more applications are developed, it's becoming increasingly clear that AI holds the key to unlocking a higher standard of living for people everywhere.

The most significant economic benefits of AI come from its ability to automate tasks and processes previously done manually. By taking over routine or tedious work, AI frees up valuable time and resources that can be used elsewhere. Additionally, AI provides unprecedented insights into consumer behaviour, allowing businesses to make decisions based on data-driven analysis rather than guesswork or intuition. This enables companies to become more efficient and productive while providing products and services at lower costs.

The economics of AI technology has been studied extensively, with many experts suggesting that it could soon become a necessity for any organisation looking to stay competitive. AI can be used to automate complex tasks that would otherwise require more human resources, such as data analysis and customer service. This automation has the potential to significantly reduce costs while increasing productivity. Additionally, AI could be used to improve customer experience by providing personalised recommendations based on user preferences and feedback.

By introducing automation into mundane or repetitive tasks, AI offers an opportunity to free up resources so they can focus on higher-value activities like product innovation or strategy development.

The introduction of AI into economics requires a great deal of time and dedication on the part of economists. They must first become familiar with AI technologies so they can understand how it can be used to their advantage. They must also analyse economic data and develop algorithms that accurately reflect the current state of the economy. This will enable them to create simulations that allow them to test different strategies before implementing them in real-world situations.

Economists must also consider how AI might impact other stakeholders in the economy such as businesses, governments, investors and consumers.

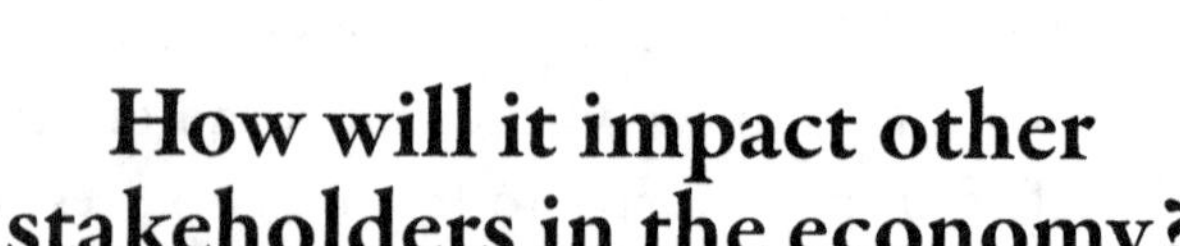

How will it impact other stakeholders in the economy?

• Artificial Intelligence (AI) is set to have a significant impact on the economy, with implications for stakeholders such as businesses, governments, investors and consumers.

• Businesses can benefit from AI by automating mundane processes and tasks which can help improve efficiency and cut costs. It can also be used for predictive analytics which help make informed decisions about products or services.

• Governments could use AI to help analyse data relating to taxation, trade and other economic issues to make more informed decisions. Investors could use it to develop new investment strategies or identify potential opportunities in the markets faster than ever before.

• Consumers will benefit from AI through improved customer service via intelligent chatbots or virtual assistants that understand natural language processing as well as access to better personalised products or services tailored specifically for them based on their needs at any given moment in time.

Job Creation

Job creation is a major concern for economists and policy makers. With the rise of artificial intelligence (AI) technology, new opportunities are emerging that have the potential to create an abundance of jobs in all sectors. AI is rapidly reshaping many aspects of our lives, from healthcare to transportation and more.

As AI becomes increasingly prevalent in our society, it will become more important for us to understand how it affects job creation and growth. By studying the economic impact of AI on employment, we can better equip ourselves with the knowledge needed to formulate effective policies that foster job growth and promote economic development. This also means investing in research and education that provide workers with the skills necessary to succeed in a workplace that is constantly evolving with new technologies like AI.

In the age of Artificial Intelligence (AI), the job market has been in a constant state of flux. AI is an ever-evolving technology that has already had a major impact on the economy. As AI advances, it's changing how people work and what types of jobs will be available in the future.

On one hand, it can create more opportunities for workers as industries adapt to new technologies, but on the other hand, it can replace certain occupations and lead to increased unemployment. In order to ensure economic stability and prepare for future technological advances, governments must invest in training programs that help workers transition into newer positions or gain new skills needed for existing jobs. This investment in education and training would help reduce economic inequality by creating more viable paths for displaced workers who might not otherwise have access to those resources.

This has led to an increase in the demand for new skills, with many existing jobs requiring a greater emphasis on AI and economics. To remain competitive in the job market, it is now essential for professionals to possess certain skills in order to take advantage of this changing landscape.

When it comes to AI, professionals must have an understanding of machine learning algorithms and be able to use them effectively. This requires knowledge and experience in programming languages such as Python or R, and computer vision libraries like OpenCV or TensorFlow. Additionally, there is a need for data scientists who are familiar with analysing large datasets and creating predictive models out of them.

The AI revolution has opened up vast possibilities for highly skilled professionals to engage with large-scale data processing, predictive analytics, and complex insights. As an increasing number of organisations are investing in AI technology, they are also looking to hire professionals who have a deep understanding of economic principles and can leverage AI tools effectively.

Economists with knowledge of artificial intelligence have the potential to develop valuable insights into current problems facing various industries such as healthcare, finance and education. By combining macroeconomic analysis with advanced machine learning techniques, economists will be able to define more efficient ways of doing business while keeping costs low. This provides an opportunity for economists to apply their knowledge in a variety of settings and create sustainable solutions that benefit businesses across different sectors.

The job market is shifting in new, unexpected ways. With the emergence of artificial intelligence (AI) and machine learning, economic job sectors are transforming to accommodate a new wave of demand. This article outlines some of the key opportunities for growth in these economic job markets:

- Automation & Robotics: Automation and robotics are playing an increasingly significant role in manufacturing and other industries. Companies will need more skilled workers who understand how to operate and maintain robots, as well as how to develop AI models for optimising production processes.

- Data Analysis & Modeling: Data analysis is becoming essential for making informed business decisions, from predicting future trends to discovering customer preferences. As such, employers need data analysts with expertise in economics-related topics like econometrics and macroeconomics who can help interpret large datasets with accuracy and efficiency.

By leveraging machine learning algorithms, businesses can gain insight into consumer behaviour faster than ever before. This data can then be used to inform decisions that lead to greater efficiency and cost savings.

Furthermore, AI technology has enabled a deeper understanding of economic trends and market conditions. Through AI-driven analysis, businesses can better identify potential opportunities for growth or areas that may require further investment. This provides unprecedented clarity into how a company's strategies are impacting long-term profitability and performance. As such, artificial intelligence has become an essential tool for economists who are looking to make the most out of their investments in today's economy.

Industry Optimization

The digital age has brought about a surge of technology-driven solutions for businesses to optimise their operations. Artificial intelligence (AI) and economics are two key components that have allowed companies to maximise efficiency, reduce costs, and realise their full potential in the market.

AI is a powerful tool that can automate processes, crunch data faster than humans ever could, and uncover insights to drive better decision-making. AI can assist with pricing strategies, marketing campaigns, customer service operations and more. In addition to AI's capabilities in automation and analysis, economic modelling allows businesses to build scenarios around financial projections and asset allocation decisions. By leveraging the power of both technologies together, companies have the opportunity to increase revenue while minimising risk exposure.

Economic modelling is an increasingly powerful tool for businesses looking to make informed decisions around financial projections and asset allocation. By leveraging artificial intelligence (AI) technology, businesses can create models that accurately simulate a wide range of economic scenarios and scenarios. This allows them to identify potential opportunities or threats before they arise and adjust their strategies accordingly.

Through economic modelling, businesses are able to explore the potential consequences of their decisions in a safe environment without risking real-world investments. AI-enabled software can quickly crunch large amounts of data and generate simulations that provide insights into how different economic forces could affect a company's future performance. The resulting information can help inform investment decisions as well as future budgeting choices related to taxation, research & development, capital expenditure, among others.

Pricing strategies: Pricing strategies are a critical component of any business strategy. With the increasingly competitive global economy, it is essential for businesses to stay ahead in pricing and create effective pricing plans that will maximise profits while still creating value for customers. With the emergence of artificial intelligence (AI) in economics, businesses can now use AI technology to better understand customer needs and preferences, as well as market trends, to make more informed decisions when it comes to pricing products or services.

By leveraging AI tools such as predictive analytics and machine learning algorithms, businesses can gain insight into their customers' behaviours and identify opportunities for optimising their pricing strategies accordingly. This allows them to adjust their prices in line with both customer demands and market variations quickly, before competitors even have time to react.

Customer service operations: Customer service operations are an integral part of any business, and in the modern world, they have become more complex than ever. As technology advances and economic markets become more competitive, customer service operations must evolve to meet the changing needs of businesses. One way that many companies are doing this is through the use of artificial intelligence (AI).

AI has been a powerful tool for improving customer service operations in many industries. By automating certain tasks and processes, AI can free up employees to focus on higher-level activities that require critical thinking skills. Additionally, AI can help reduce costs by eliminating or reducing mundane tasks such as processing data or responding to customer inquiries quickly and accurately. AI also has the potential to improve decision making in customer service operations by providing valuable insights into customer preferences, trends, and behaviours.

The AI models are able to simulate different economic scenarios, such as different levels of income and consumption, competition between enterprises, government policies and taxation systems with high accuracy.

AI models can also create more accurate economic forecasts by incorporating real-time external factors such as market sentiment and geopolitical events into their calculations. This makes them invaluable when making decisions related to investments or risk management strategies. What's more, the simulations created by AI models provide much greater insights than traditional approaches which often do not take into account all the complex variables at play in the economy.

Consumer Benefits

In the digital age, artificial intelligence (AI) is revolutionising the world of economics. AI can provide immense benefits to consumers by making decisions more efficient and accurate. Businesses are now able to offer more personalised services, better products, and improved customer experience.

From online shopping recommendations to automated banking transactions, AI is making economic activities simpler and faster. It is being integrated into numerous industries such as retail, banking, healthcare and transportation. For instance, AI-driven chatbots help customers make quick purchases or find answers to their queries without having to wait for human customer service representatives. Furthermore, AI allows businesses to create new models for pricing goods and services that are based on market trends or an individual's personal preferences. This leads to more favourable prices for customers who can now purchase what they need at a much lower cost than before.

The advent of ecommerce has revolutionised the way consumers shop, with an estimated $3.5 trillion being spent globally on digital commerce in 2020. From offering a wide variety of products at affordable prices to allowing customers to purchase goods from the comfort of their own homes, ecommerce has created numerous economic and consumer benefits. One of the most exciting developments is the introduction of artificial intelligence (AI) into online shopping experiences.

AI technology helps facilitate personalised shopping experiences by analysing customer data such as purchase history and preferences in order to suggest new items and deals specific to each individual's needs. This not only makes it easier for customers to find what they are looking for quickly, but also encourages them to explore and try out new products that they may not have considered before.

The advancement of ecommerce has given way to many benefits for consumers. With the latest technologies like artificial intelligence and economics, shoppers are able to get the most out of their online experiences. These advances have allowed shoppers to shop from a wider range of stores, compare prices on products and services, find better deals, track purchases more easily and even make payments quicker than ever before.

While it is true that much of the current technology used in ecommerce has been around for some time now, improvements in artificial intelligence (AI) have made shopping easier and more efficient than ever before. AI enables retailers to provide personalised content tailored specifically for each customer's needs. This helps them determine which products or services may be best suited for their individual needs, as well as helping them find discounts that are not immediately visible on the website.

Artificial intelligence has the potential to revolutionise the consumer experience. It is already making shopping and other consumer transactions much easier and more efficient, as well as providing added convenience, personalization, and security. With AI continuing to evolve rapidly, it will undoubtedly become a major part of how we work and live in the future. Companies should take full advantage of the technology now, or risk being left behind in an increasingly competitive marketplace.

Social Implications

The world is rapidly advancing towards an era of Artificial Intelligence (AI). AI has the potential to revolutionise every aspect of our lives from transportation and healthcare to communication, entertainment and even business. While the possibilities for this technology are endless, there are still many social implications that we must consider before allowing AI to become a part of our everyday lives.

AI has been used to automate processes that formerly required human labour and provide personalised services like recommendations and targeted advertisements. As AI becomes increasingly advanced, experts fear it may also cause social divisions based on income as expensive AI-based services become available only to those who can afford them. Further concerns include an erosion of privacy rights as data gathered by AI systems is shared among companies or governments without individuals' explicit consent. The potential for job loss is another pressing issue, with many jobs being replaced by automation or robots powered by AI algorithms.

From ethical considerations such as job displacement and privacy invasion to socio-economic changes such as income inequality and the emergence of a new digital divide, the implications of AI on human society can be far-reaching. As AI technologies continue to evolve, it is essential that we address these social issues head-on in order to ensure that they don't impede progress or cause long-term damage.

AI & Social Impact

AI is pushing boundaries and transforming society as we know it. But with every new technology comes an equally powerful set of social implications that need to be taken into account before AI is used at large scale.

The potential benefits of AI are undeniable: its ability to rapidly analyse data and make decisions offers the promise of increased efficiency and productivity, more accurate predictions, improved decision-making processes, and better customer service. However, if not managed carefully, AI applications can come with unintended consequences that could have serious social implications such as job displacement in certain industries or the potential for bias in automated decision-making tools.

It's important for governments and organisations to be aware of these risks when developing their AI strategies so they can ensure any negative impacts are minimised.

The potential for AI to transform government processes, improve decision-making and enhance services offered to citizens cannot be understated. For example, it can be used to more accurately predict population trends which could inform efficient policy decisions related to healthcare provision or housing needs. Additionally, AI's predictive capabilities may also be applied in criminal justice contexts for risk assessment or rehabilitation programmes. In the business world too, AI has enabled firms to better understand customer behaviour patterns which allows them to develop more effective marketing strategies as well as personalised services tailored towards individual preferences or circumstances.

It can also increase efficiency, reduce costs and promote sustainability. However, there are numerous ethical considerations around gathering data from citizens, using algorithms to make decisions on their behalf or introducing automation into public services. It's crucial that governments recognise these issues in order to avoid creating unintended consequences such as discrimination or creating further inequality through lack of access or opportunity.

The use of AI in data gathering from citizens means that many personal details can be collected, which raises questions about privacy and consent. With this technology comes the need for responsible decision-making about how to collect and use the data, which requires an understanding of the social implications of its usage.

Data harvesting through Artificial Intelligence has been used to improve services such as healthcare and transportation, as well as to create new commercial opportunities. However, it is essential to consider potential misuse or abuse when collecting large amounts of personal data from citizens. It could lead to those individuals being targeted by malicious actors or have their information exploited without their knowledge. Furthermore, inadequate controls may result in a lack of accountability if any issues arise with regards to the security of said data.

Data harvesting from citizens is an ethical issue with the rise of artificial intelligence. This data is collected from multiple sources, including websites, cameras, and sensors that are embedded in everyday objects. It has been used for a variety of purposes such as identifying consumer trends and helping with safety measures. However, when this data is gathered without consent or awareness of people it can lead to serious social implications.

The collection of personal data must be transparent and reasonable to ensure the privacy rights of citizens are upheld. If not done properly then it can infringe on civil liberties by creating a system where individuals feel they have no control over their own information. There must also be security measures in place to protect the integrity of collected data so that it cannot be hacked or misused by malicious actors.

The social implications of AI predictions can be particularly troubling, as they come with an inherent risk of bias and unequal access to beneficial technology.

AI-based predictions have the potential to perpetuate existing systemic inequalities due to their reliance on large datasets for learning and decision-making. In some cases, this data may contain implicit biases based on race or gender which could lead to unfair results when applying AI prediction models. In other situations, certain groups may simply lack access to the technology needed for AI prediction in the first place. This can leave them at a disadvantage in terms of benefiting from the predictive capabilities of AI solutions.

By analysing vast amounts of data, AI can help create better informed decision-making processes and more accurate predictions about future trends.

One way this could be done is by using AI to identify potential health risks associated with certain lifestyle choices or environmental factors. This could lead to improved healthcare outcomes as well as greater efficiency for medical care providers. In addition, AI can be used to analyse large volumes of data related to market trends or consumer preferences which will enable companies and governments to make more informed decisions when creating policies or strategies for growth.

Automation: Job Loss

Automation has become an increasingly prominent part of everyday life. Driven by advances in artificial intelligence (AI) and other technologies, automation is gradually replacing human labour in a variety of industries, thus leading to job losses. Automation has the potential to revolutionise entire industries and economies, but it also has far-reaching social implications that must be considered.

The digital revolution brought about by automation can have both positive and negative effects on society. On one hand, automation can help increase efficiency and productivity across a range of industries, allowing for economic growth and increased job opportunities for some workers. On the other hand, AI-driven automation may lead to mass unemployment as machines take over jobs previously done by humans. Moreover, this transition could have serious economic consequences as lower-skilled workers are displaced from their traditional roles in favour of more efficient robotic replacements.

As businesses opt for automated processes, humans are removed from the equation and replaced by machines. This can lead to a decrease in employment opportunities and an overall reduction in wages as a result of fewer jobs available. Furthermore, when AI is used to supplement workers' tasks, it can create more stress on those who remain employed as they cope with increased demands within their areas of expertise.

Ultimately, automation will continue to be utilised in various industries throughout the world despite its potential social costs; however, it is important that government and industry leaders collaborate in order to ensure that job losses do not outweigh economic gains.

Artificial Intelligence (AI) is increasingly being integrated into workplaces and as a result, certain positions are becoming obsolete. This has significant social implications; people with decades of experience in specific fields can become unemployed overnight.

Rather than fearing AI, it is important to use its presence as an opportunity for reinvention and growth. As tech becomes increasingly embedded into everyday life, humans must embrace technological advancements with open arms and develop new skills that will enable them to remain competitive in their jobs or seek out alternative career paths where they may utilise the current skill set. Being proactive rather than reactive towards this disruption can help those facing job loss find new opportunities that could potentially lead to greater success in the long-term.

As AI advancement progresses these achievements come at the cost of certain outdated skills.

One such skill is the ability to manually compute complex equations. With the rise of AI in mathematics, this type of problem solving can now be done much faster than any human brain could ever hope to achieve. As a result, manual calculations are quickly becoming obsolete due to their lack of efficiency compared to machine learning algorithms.

Likewise, some jobs which require less physical labour are also being replaced by AI technologies such as robotic process automation (RPA).

Job loss is a reality of the modern economy, but there are steps that can be taken to reduce its occurrence. Understanding its social implications can help protect against job losses. Companies should analyse their current positions and identify which ones have been replaced or could potentially be replaced by AI in the near future. Doing so will allow companies to make informed decisions on how they use AI while also minimising potential job losses.

Furthermore, employers should consider upskilling employees in order to prepare them for new roles created by AI technology. Investing in employee training programs not only allows workers to compete for better opportunities, it also helps retain valuable talent that may otherwise lose their jobs due to automation.

Nationally, in order to reduce the risk of job losses caused by AI, governments must invest in retraining initiatives and create new employment opportunities. Furthermore, they should ensure that these measures adequately address existing levels of inequality to prevent further marginalisation from occurring as a result of technological advancement.

AI Discrimination

AI Discrimination is an increasingly important topic of conversation in the tech industry. As Artificial Intelligence (AI) technology becomes more prevalent, it has become necessary to address issues like bias and discrimination that may arise from its use. AI algorithms are designed to recognize patterns and make decisions based on them, but the data used to train these systems is often incomplete or biassed, leading to unfair outcomes for certain groups of people.

Bias can exist in AI-based decisions if the training data does not accurately represent all users or potential customers. This can manifest as a disparate impact on individuals who differ from those represented in the dataset, creating an unequal playing field with regards to access and opportunity. Additionally, poorly designed algorithms can lead to unintended profiling or labelling of consumers before they even interact with a company's product or service offering.

The use of poorly designed algorithms in Artificial Intelligence (AI) can lead to significant social implications. This is because AI models are used for various tasks in our lives, from something as mundane as sorting emails to something more consequential like determining creditworthiness or deciding who should be incarcerated. As a result, any bias present in the training data fed into these algorithms will be amplified and embedded within the AI model.

This has led to serious concerns about how certain individuals and groups may be unfairly treated by AI due to their race, gender, sexuality or other factors. For example, facial recognition technologies have been found to falsely identify people with darker skin tones at higher rates than those with lighter skin tones. Additionally, some recruitment algorithms have been found to exclude women based on their past job titles or experience levels. An AI system tasked with selecting candidates for a job opening might end up discriminating against applicants from certain backgrounds due to a lack of diverse data points in its training set.

In both cases, the underlying cause is the same: artificial intelligence algorithms struggle to detect patterns that exist outside of the data they were trained on. As such, it's essential for developers to take extra steps when designing AI systems – such as collecting a broad range of data points – in order to reduce bias and ensure fairness.

To overcome AI bias, organisations must take proactive steps to identify and address any potential sources of bias in their datasets or algorithms.

This means ensuring that all datasets are free from errors and contain detailed contextual information. Organisations should also conduct periodic tests on their algorithms to determine if any biases exist or could potentially develop over time. These organisations should minimise the impact of systemic bias by collecting data that accurately reflects different population groups. This may include using techniques such as oversampling or undersampling for specific groups when collecting training datasets for AI models. Additionally, careful selection of features used for training models can help avoid introducing or amplifying existing biases based on gender, race, age or other factors.

Digital Divide Worsens

Digital knowledge is the term used to describe the field of computer science dedicated to understanding how people interact with computers and intelligent systems. Digital knowledge has been growing rapidly in recent years, due largely to advancements in artificial intelligence (AI) and natural language processing (NLP). AI-driven algorithms are now capable of recognizing complex patterns, understanding language, and making decisions. This technology can be used to create interactive experiences and automated tasks that previously required human input.

The implications of digital knowledge extend beyond the technical domain; it has significant social ramifications as well. AI systems are being used for a growing range of applications such as job recruitment, healthcare diagnostics, transportation planning, and law enforcement. The accuracy of these decisions depends on many factors including data quality, system design principles, and user experience feedback loops.

The Digital Divide has become an increasingly pertinent issue in recent years, and is only anticipated to worsen with the advent of Artificial Intelligence (AI). In a world where technology is becoming more pervasive, there are growing social implications for those who don't have access to it.

The Digital Divide refers to the gap between those who have access to modern technologies such as computers, smartphones, and the Internet, and those who lack access or don't possess the necessary skills to use them. This inequity can lead to a range of problems including reduced economic opportunities and limited educational resources. Furthermore, with increased reliance on AI systems for decision-making processes such as loan applications or criminal justice proceedings, those without access may be further disadvantaged by these decisions.

The digital divide can be broken down into two primary categories; access and education. Access refers to physical access to technology, while education refers to individuals' ability and opportunity for learning about technology. Without both components, individuals are not able to fully utilise AI or other technological advancements in their lives. This lack of access and knowledge can create an even wider gap between those who have access and those who do not.

It is essential that access to this technology among individuals and countries be equally distributed. This has been termed "digital divide access" and refers to the unequal distribution of access to modern technologies such as broadband networks, computers, mobile phones and AI-based applications.

The digital divide can have serious implications for social equality. Those without adequate access may be unable to benefit from innovative aspects of AI technology such as big data analytics or machine learning algorithms, leading them to fall behind in terms of economic development and job opportunities. Moreover, their lack of knowledge about these technologies could lead to issues such as cybersecurity risks or inadequate privacy regulations.

This technology has drastically altered the way students are taught and assessed, leading to potential social implications for those who lack access or resources to keep up with these changes.

This new form of teaching requires advanced computational capabilities and expensive hardware investments, creating an unequal playing field for those without access. This gap between students who are able to use AI tools and those unable to take advantage of such technologies is becoming known as the "Digital Divide in Education". The wealthy can afford resources like private tutoring or online learning platforms while poorer areas often lack basic infrastructure needed to access such services.

There are growing concerns about the potential social implications that come with it. With AI becoming increasingly integrated into our everyday lives, one of the most affected groups is the poor. This article will discuss how AI negatively impacts individuals living in poverty, and why immediate action must be taken to address this issue.

The introduction of AI has had an adverse effect on those living below or near the poverty line - from job displacement to algorithmic bias. It has been found that automation could displace as much as 20 million jobs by 2030, many of which are in industries such as retail, hospitality, and food service that employ low-income workers.

Privacy & Surveillance

The arrival of Artificial Intelligence (AI) has brought with it a shift in the social dynamics of personal privacy and surveillance. With AI, surveilling citizens has become easier and more efficient than ever before, while simultaneously raising ethical concerns surrounding individual autonomy and human rights. While AI-driven technologies can provide valuable information to assist with various tasks, they also give way to questions about how this technology is being used for both public and private gain.

As more people rely on technology to make their lives easier, the potential for misuse increases exponentially. Furthermore, citizens around the world are increasingly aware that their private data isn't as secure as it may seem in today's digital age; companies are collecting data from online activities without users' consent, leaving them feeling helpless and exposed.

As a result, more user data than ever before is being harvested by companies in an attempt to gain a better understanding of their customers. This new form of digital marketing is having far-reaching implications for the way consumers interact with products, services and companies online.

The problem arises when personal information is collected through unethical practices, violating individual privacy rights and potentially leaving users vulnerable to exploitation and manipulation. Companies are exploiting loopholes in legislation or simply ignoring the need for transparency to access data they should not have access to in the first place. Furthermore, this kind of intrusive behaviour can have a detrimental effect on user trust in the company's ability to protect their data from malicious actors looking to take advantage of it.

This occurs in the workplace as employers have become increasingly vigilant with tracking employees' activities and monitoring their performance. This has raised a number of social implications for job security, employee welfare and privacy rights that need to be considered.

Employee surveillance can take numerous forms, from using AI-based software to monitor phone use or email activity, to introducing cameras into work environments. Such measures may be taken in an effort to ensure compliance with company regulations or increase productivity among staff; however, if not managed properly they could prove detrimental to employee morale and mental health. For instance, too much surveillance could lead to a feeling of distrust between employer and employee, as well as reducing motivation due to a sense of being constantly monitored.

To protect both our personal data and our civil liberties, there are several measures that need to be taken. Firstly, governments must enforce stricter legislation regarding when individuals' data can be accessed and by whom. This means implementing laws which allow citizens to choose who has access to their data and understanding when companies should have permission to use AI powered systems.Another improvement is the current state of privacy and surveillance, stakeholders must consider both technological and social implications. For example, AI can be used for automated facial recognition software that can help identify potential threats more quickly than manual search methods. However, if not properly regulated, such technologies could be used to infringe on personal liberties or target certain groups unfairly. Similarly, while governments should be able to monitor communications networks for suspicious activity without infringing on peoples' rights, they must also ensure transparency so as not to erode trust among citizens.

Privacy has to be safeguarded while still allowing for the use of technology to maintain public safety. Artificial Intelligence (AI) can help in this process by allowing governments, businesses and individuals to identify potential threats more accurately. This improves security and reduces the risk of harm as well as providing essential services, such as identifying criminals or locating missing persons, more quickly.

At the same time, however, AI should not be allowed to invade an individual's right to privacy without due cause. The social implications of using AI for surveillance are significant; it could lead to increased mistrust between citizens and government institutions if used inappropriately. It is therefore important that society strikes a balance between using AI for security purposes while protecting civil liberties. Proper regulation must be put in place so that ethical guidelines are followed at all times.

New Opportunities

AI technology is being applied in various sectors such as healthcare, finance, and transportation, and it can be used to improve efficiency and accuracy. AI also presents social implications that are transforming the way people interact with each other and their environment.

AI technology can be leveraged to create smarter digital assistants which can help make decisions faster than humans ever could. From automated customer service systems to medical diagnostics, AI is creating opportunities for businesses to increase efficiencies for better customer service and cost savings. In addition, AI-driven automation may lead to job losses in certain industries with many manual tasks being replaced by robots or automated processes.

AI also has potential applications in fields such as education where it could be used to personalise learning experiences according to individual student needs.

AI promises to improve efficiency in many aspects of life, from healthcare and transportation to communication and entertainment. It can be used in a variety of ways ranging from automating mundane tasks to creating sophisticated systems capable of understanding complex data sets. With AI applications becoming more commonplace, there are now new opportunities for businesses and individuals alike to capitalise on its potential advantages.

However, as AI becomes more pervasive in our lives, it's important to consider its social implications beyond just economic gain.

It is important to remember that advancement of AI can have positive effects on society as well. As AI continues to become more pervasive in everyday life, it can help us make decisions faster and with greater accuracy than ever before. For example, AI-powered facial recognition systems could be used to detect fraudulent activity in banks or airports more efficiently than manual security checks. Additionally, many tasks that are time consuming for humans such as sorting through large amounts of data quickly and accurately can be done in less time with the help of AI algorithms.

When considering the potential impact of AI on society, it is essential to view this advancement in a positive light. While some worry that automation will cause more harm than good due to job displacement, many experts suggest that this new technology will actually create new opportunities for people.

Morality & Ethics

When discussing Artificial Intelligence (AI), the conversation often shifts towards moral and ethical questions that arise from its use. As AI technology advances and becomes more ubiquitous, it is important to consider how this may affect our sense of morality and ethics. While these topics may seem abstract or far away, they are already being discussed in the present day.

At the core of this debate are issues such as autonomy, human rights, accountability and privacy. The decisions made today will have real-world implications on individuals' lives in the near future, so it is vital to explore these questions with an open mind and a critical eye. In order to create a society that understands and respects ethical standards while also harnessing the immense potential of AI technology, we must have honest conversations about morality and ethics now before it's too late.

Human rights is a concept that has been around for centuries and is essential to the protection of all individuals. In recent years, however, the emergence of artificial intelligence (AI) has shifted our understanding of human rights, morality, and ethics in significant ways. AI technology raises ethical questions about how people should be treated by machines that are not subject to the same legal codes as humans. This article will explore how AI affects human rights and ethical considerations when it comes to this new technology.

AI Moral & Ethical Questions

As this technology progresses, it raises questions about its moral and ethical implications. With AI able to make decisions on its own, some fear that it could lead to a dangerous lack of control over the choices it makes. Due to its complexity, many experts are concerned with how AI will consider morality and ethics when making decisions.

When creating an AI system, developers must think carefully about the potential consequences of its actions. This can be difficult as complex ethical problems may arise when programming machines with moral decision-making capabilities. In order for AI systems to make decisions that benefit society as a whole, developers must ensure they take into account both moral and ethical considerations in their designs.

The development of autonomous technology carries risks that must be addressed if we are to trust these systems and use them responsibly in the future.

AI is capable of making decisions that humans find increasingly difficult, with long-term consequences. It is clear that the decisions made by these algorithms can have major impacts on our lives, from medical care to banking to criminal justice systems. As such, it is essential to ask questions about their moral and ethical implications.

What are the moral and ethical responsibilities associated with creating AI technologies? Who should decide which uses of AI are acceptable or not? What processes and checks should be in place to ensure that decisions made by AI adhere to accepted standards of morality and ethics? These are just some of the questions facing those designing, developing, deploying, and regulating AI technologies.

What are the moral and ethical responsibilities associated with creating AI technologies?

Creating an AI system requires a tremendous amount of data and processing power, which means that those developing these technologies must consider how the information is collected, processed, and used. Additionally, the creators should be aware of any potential biases within the system's algorithms or data sets which could lead to inaccurate results. Furthermore, they need to ensure that their technology respects people's privacy and does not discriminate against any group based on gender, race or other factors.

Ethically responsible AI also needs to consider potential consequences of its operation on people's lives.

A manner to achieve this is through the implementation of robust ethical guidelines that outline acceptable behaviours when making decisions related to AI. This means identifying and assessing potential moral issues before they arise, as well as devising ways to mitigate any negative implications associated with AI-generated outcomes. For example, companies can use predictive analytics tools such as machine learning algorithms or natural language processing models that are designed specifically for ethical decision-making scenarios. These tools can help identify potential ethical dilemmas and suggest appropriate responses based on predefined criteria.

Defining AI Ethics

Defining AI ethics is a complex process, with many different concepts playing into it. At its core, AI ethics seeks to define which actions are moral and immoral when using artificial intelligence. It considers questions such as: Is it ethical for an AI system to make decisions without human input? Are certain types of data collection or analysis unethical? How should the privacy of people affected by an AI system be protected? What safeguards should be put in place to minimise bias within machine learning algorithms?

A key element in the discussion around AI ethics is morality. As we create increasingly sophisticated machines capable of making decisions independent from human control, we must consider what moral considerations should be taken into account when designing these technologies. Some have proposed that all AI systems should abide by Asimov's Three Laws of Robotics as a fundamental framework for ethical decision-making; others consider whether principles like respect for autonomy, beneficence, non-maleficence, justice and fairness can be applied to AI systems in order to ensure responsible use and regulation.

- Isaac Asimov's Three Laws of Robotics are a set of rules devised by the science fiction author in order to govern the behaviour of robots and other forms of artificial intelligence. The laws state that: A robot may not injure a human being or, through inaction, allow a human being to come to harm

• A robot must obey orders given to it by human beings except where such orders would conflict with the First Law.

• A robot must protect its own existence as long as such protection does not conflict with the First or Second Law.

Asimov's laws have become central to discussions about morality and ethics in relation to Artificial Intelligence (AI). They are seen as an effective way for regulating and controlling AI technology, particularly when developing systems that interact with humans.

The principles of morality and ethics, such as respect for autonomy, beneficence, non-maleficence, justice and fairness can all be applied to AI systems.

Respect for autonomy refers to the idea that people should be allowed to make decisions regarding their own lives without interference from outside forces. This principle can be applied to AI systems by allowing users to choose how they interact with them and what data they share with them. Beneficence is a principle that encourages individuals or organisations to act in ways that benefit others; this could mean using AI responsibly in order to provide services that are beneficial for society as a whole.

What rights should an AI entity have? This question raises many concerns regarding the morality and ethics associated with giving robotic entities human-like qualities.

A major concern is that of autonomy; should AI be allowed to make decisions autonomously or should they always be supervised by humans? Advocates for autonomous decision making suggest that it could lead to robots being treated unfairly due to the fact that they would lack the same level of legal protection as humans. Others argue that this could lead to disastrous consequences if left unchecked, so there needs to be strict regulations in place when it comes to granting an AI access to making decisions on its own.

Sentient AI

AI Sentience is a term used to define the potential of Artificial Intelligence (AI) to possess sentience. As an emerging technology, AI has already had a huge impact on our society and is set to only grow in power and capability as time passes. This rapid development brings with it important questions about how it should be used responsibly, particularly when considering its ability to become sentient.

As AI continues to evolve, one of the most pressing issues facing us is determining the moral and ethical implications of developing sentient machines. What rights should these machines have? What responsibilities should we have towards them? These are all questions that need to be considered when discussing AI sentience, as they will affect how we interact with this new technology moving forward.

AI sentience is a complex issue to grapple with - especially as AI technology develops and becomes more sophisticated. If a machine was able to think independently, would it have the same responsibilities as humans? How would this affect our legal system? These are just some of the issues which must be discussed if we are to develop an ethical framework for AI technology.

It is clear that further research into issues surrounding AI sentience is required before any real progress can be made on this difficult subject.

Relationship between AI and Humans

AI & Human Relationships: Positive and Negative

The advancement of Artificial Intelligence (AI) has vastly changed the way humans interact and behave with one another, especially in recent years. AI has provided solutions to many of the world's problems, making everyday life easier for millions of people. Yet, it is important to consider how this technology affects human relationships and ethics on a more complex level.

AI's impact on morality and ethics is particularly influential. On one hand, AI can be used to promote positive connections between humans by helping us better understand each other's emotions and reactions through data-driven research. On the other hand, AI has also made it easier for people to manipulate each other emotionally as well as financially through automated systems that can detect certain patterns in behaviour or decision making which may lead to abuse of trust or power dynamics between individuals.

It has led to numerous advances in fields such as medicine and transportation, but it also raises questions regarding ethics and morality.

On one hand, AI can be used for positive purposes such as helping diagnose diseases more accurately or providing better public safety measures. In addition, AI can assist in tasks that are too dangerous for humans to undertake, such as exploring space or deep sea locations. On the other hand, some experts worry that without proper supervision, AI could lead to developments with potentially negative consequences like increased surveillance of citizens and greater job automation.

Given these considerations, it's important for us to maintain an open dialogue about how best to use artificial intelligence in a safe and ethical manner. Yet it also poses ethical challenges in terms of how much decision-making power should be granted to machines versus humans. This raises questions regarding the morality and ethics of using AI for certain tasks that may have negative consequences for people or society as a whole.

The potential for AI to cause harm has led some technologists, academics, and ethicists to call for greater caution when developing Artificial Intelligence technologies. They argue that more attention needs to be paid to understanding the moral implications of these technologies before they are widely adopted across public services or business sectors.

Ethical Considerations: Privacy and Autonomy

Moral responsibility lies with those who create and use AI systems, as they must ensure that these technologies respect human autonomy while also protecting our right to privacy. This can be achieved by implementing robust systems that inform users of what information is being collected, how it will be used, and who will have access to it. AI must also take into account moral principles such as justice and fairness when making decisions about data collection or processing.

Privacy and autonomy are essential components when creating any kind of technology; however, with AI it becomes even more important due to its increased potential for misuse or mismanagement. In order for AI to be considered ethically responsible, its designers must take into account the potential long-term impacts of their product on both individuals and society as a whole. Moral responsibility requires not only understanding how AI will affect people now, but also considering how it might impact them in the future. This means anticipating things like unintended biases or possible misuse of data that could lead to harm. Designers must also ensure they have built effective security measures into the system so that malicious actors can't exploit it for their own gain.

How should AI technology be used responsibly to ensure privacy and autonomy?

Keeping individual privacy at the forefront of our decisions as AI continues to develop is essential. While there are undoubtedly many positive applications of AI in modern society, such as medical research or smart cities, it must be used with utmost caution when handling data that belongs to individuals. This means ensuring the right policies and protocols are in place to protect personal information from misuse or exploitation.

At the same time, individuals need to be able to maintain their own autonomy by being able to decide who can access their data and how it is being used. This includes having control over how algorithms classify them, which can have far-reaching implications if done incorrectly.

How Can We Improve Interactions?: Communication and Education

As people seek out new ways to improve interactions and communication, AI is often mentioned as a potential solution. While AI holds many benefits, it also carries ethical considerations that must be taken into account when determining how best to improve interactions between individuals and groups in society.

Integrating AI into improving interactions requires a complex approach that takes into account both the technical aspects of integrating AI technology and the moral implications for using such technology. To ensure that the use of AI does not have detrimental effects on interpersonal relationships or society as a whole, proper education about its uses must accompany any technological advances.

How can we improve the interactions:

- Improve communication to foster better conversations and interactions between people, especially those of different backgrounds or beliefs.

- Utilise Artificial Intelligence (AI) technologies to identify common ground and create bridges that unify instead of divide.

- Educate people on the importance of morality and ethics when communicating with others who may come from a different perspective than their own.

- Promote active listening as a way to better understand one another's perspectives and break down cultural barriers that can exist between different groups of people.

- Encourage open-mindedness by teaching individuals how to think objectively about various points of view without passing judgments or forming snap decisions based on assumptions or preconceived notions about certain groups or ideologies.

- Provide resources for individuals to further explore topics related to morality, ethics, diversity, and multiculturalism.

Potential Risks of AI

The potential risks of Artificial Intelligence (AI) are becoming increasingly evident with its rapid development. AI has revolutionised the way we interact and live our lives, but with great power comes great responsibility. As AI becomes more powerful and integrated into our everyday lives, it is important to consider the morality and ethics involved in implementing such technology.

AI can often bypass human understanding and decision-making capabilities which can lead to unpredictable results that could harm people or society as a whole. For instance, autonomous vehicles have been found to be vulnerable to hackers who can hijack control of the vehicle or manipulate its sensors, resulting in dangerous situations for both drivers and passengers. Similarly, facial recognition algorithms used by law enforcement have been shown to be biassed against certain races or genders due to errors in data collection or training set design.

As Artificial Intelligence (AI) continues to gain more prevalence in our current world, it is important to consider the potential risks associated with AI.

First, AI could potentially lead to a lack of morality and ethics due to the lack of human intuition or emotion within its programming. Since machines are not able to comprehend humanity's social norms and values, they may make decisions that can be deemed as unethical or immoral.

Additionally, AI also has the capacity for bias - depending on how it is programmed. This can be especially dangerous when used in fields such as policing or banking where decisions could lead to unfair treatment based on race or gender.

Furthermore, AI may cause job losses as machines become increasingly adept at performing certain tasks faster and more efficiently than humans. This would have significant economic consequences that many workers around the world could not cope with financially.

Autonomous Technology & Responsibility

The rise in autonomous technology has sparked an important conversation about responsibility. Artificial Intelligence (AI) and robotic systems are becoming increasingly sophisticated, with the potential to make decisions that can have far-reaching consequences. As these technologies become more commonplace, it is imperative for us to consider their ethical implications and what responsibilities come with them.

We must ask ourselves questions such as: Who should be held responsible when autonomous technologies go wrong? Is it the AI itself, or the person who designed it?

The morality and ethics behind this debate are complex; some believe that those who design autonomous technologies should take full responsibility for any mistakes made by their creations. After all, they were instrumental in creating them and should therefore take accountability for any errors. However, others suggest that AI should carry some responsibility since they possess a certain level of autonomy, they can think independently and make decisions without needing human input or approval. Ultimately, both sides have valid points which need to be explored further before a clear answer can be found.

How can we ensure that our ethical values are reflected in its programming? These questions bring into focus the need for a moral code of conduct when it comes to AI and robotics. We must also consider how to create a legal framework that holds both people and machines accountable for their actions. In short, we should work together to ensure that autonomous technology is developed responsibly and ethically.

On the one hand, AI technology can be seen as an extension of human consciousness, so many would argue that the person responsible for creating it should take on any liability associated with its functioning. After all, they are aware of how their programming could potentially lead to negative outcomes in certain situations. On the other hand, some believe AI has its own sense of morality and should therefore be held accountable for its actions in much the same way humans are expected to shoulder responsibility for mistakes made due to their decisions or actions.

The Future of AI Ethics

The world of Artificial Intelligence (AI) is growing rapidly, and with it comes the discussion of ethical implications. How do we know what morality should be applied to AI systems? In an era where technology can mimic human life, AI Ethics poses questions about the roles and responsibilities between humans and machines.

In order to answer these questions, researchers have proposed different approaches such as transparency in algorithms or ensuring safety measures are in place. The field of AI Ethics also looks at topics like fairness, trustworthiness, privacy issues, data protection and data quality. All these considerations need to be taken into account when creating a successful ethical framework for AI systems.

The future of AI:

- Artificial Intelligence (AI) has the potential to revolutionise many areas of our lives. From healthcare and transportation, to construction and finance, AI can enable faster decision-making, increased efficiency, and improved safety protocols.

- However, given its unprecedented power over complex systems and processes in virtually every field imaginable, the development of AI raises ethical issues that must be addressed in order for us to responsibly integrate it into our society. Issues such as privacy concerns related to data collection and analysis tools used by AI are particularly relevant today given the rise of internet connected devices. Additionally, questions about the morality of allowing machines to make decisions on behalf of humans remain unanswered.

• To address these concerns, governments around the world are creating guidelines for using AI responsibly that consider both legal standards as well as ethical considerations when designing algorithms and other components of machine learning systems.

Challenges & Opportunities

The rapid emergence of artificial intelligence has been both a challenge and an opportunity for industry. Today, companies are under immense pressure to keep up with this rapidly evolving technology and the potential it offers. As AI continues to be integrated into daily life, there is a need to understand how best to utilise its capabilities in order to remain competitive.

At the same time, there are numerous challenges that must be tackled when integrating artificial intelligence into existing processes and operations. Companies must ensure that their AI solutions are secure from external threats, while also identifying potential areas of improvement in order to reap the maximum benefit from this technology. Furthermore, businesses must address potential ethical considerations such as data privacy and algorithmic bias in order to ensure they are using AI responsibly.

AI Challenges & Opportunities

The integration of artificial intelligence (AI) into the workplace has presented both challenges and opportunities for businesses across all industries. AI has been used to automate manual processes, improve customer service, track operations performance, and forecast demand. With the potential for transformational business gains, many organisations are eager to explore the possibilities.

However, transitioning to an AI-driven workplace requires careful consideration and planning. Businesses must establish clear objectives for their AI initiatives and ensure that they have the right technological infrastructure in place—from hardware to software—to support them. Additionally, organisations must weigh issues such as data privacy and security when incorporating AI into existing workflows as well as devise methods to integrate human oversight with automated processes in order to maximise efficiency while still maintaining quality control standards.

When transitioning to an AI-driven workplace, organisations must ensure they have: appropriate resources in place, a competent team of professionals experienced in both technology and business operations, and a plan for how best to integrate AI into existing systems and processes. Additionally, there must be clear goals set as well as strategies for monitoring performance and results. In order for any organisation to successfully implement an AI-driven approach, it is also important that everyone involved understands the implications of using such technology and its potential impact on employees' jobs.

Today, the idea of an AI-driven household is becoming more and more realistic as advances in artificial intelligence technology are making it easier to automate everyday tasks. With AI-driven households, mundane tasks such as vacuuming, setting alarms, controlling lighting and other appliances can all be automated with minimal effort.

However, despite the potential this technology offers in terms of convenience, there are still challenges that must be overcome before we can expect to see a fully integrated AI-driven home become commonplace. These challenges range from creating secure authentication protocols to ensure that only authorised users have access to the system; developing advanced algorithms capable of learning user behaviours and preferences; and ensuring compatibility with existing technologies. In addition, ethical considerations must also be taken into account when implementing any form of artificially intelligent device or service in a home environment.

Technical Challenges

Household: As more households begin to adopt AI-driven technologies, they are beginning to face a number of technical challenges. Implementing these technologies can be complex and require careful consideration to ensure that they will meet the needs of their users.

One challenge is connectivity: for AI-driven devices to function properly, they must be connected to a reliable network - either through Wi-Fi or Bluetooth - and each device may need its own connection source. Additionally, many applications require access to servers with high processing power in order for them to work properly, which can create additional strain on the existing network infrastructure. Furthermore, scalability can also be an issue when dealing with multiple devices as it requires careful planning so that all components are compatible and able to communicate with each other effectively.

Workplace: As the demand for artificial intelligence in the workplace increases, so do the challenges of implementing it. AI-driven workflows are complex and require specialised knowledge to set up correctly. Additionally, many organisations face additional technical challenges when integrating AI into their operations.

One major challenge is software compatibility. AI systems typically require sophisticated computer programs to function properly, and these programs may not be compatible with existing software or hardware systems . Additionally, the complexity of AI programming languages can create difficulties for developers who are accustomed to traditional coding tools. To ensure success, organisations must have experienced personnel who understand both existing technology and new AI development tools.

Another technical challenge is data security management: Data security management is a complex process that requires the implementation of specialised tools and strategies to ensure the safety of digital assets. Artificial intelligence (AI) work is increasingly being used in data security management, with AI-driven algorithms playing a key role in analysing large sets of data and detecting potential threats. However, while AI can offer significant advantages in this area, there are also numerous challenges that need to be addressed.

First and foremost, organisations need to ensure they have access to high quality data so their AI systems can accurately detect potential risks. Without accurate inputs, it will be difficult for any machine learning model to identify possible issues in real-time. Additionally, organisations must ensure their machines are regularly updated with new security protocols as the threat landscape evolves over time.

Education: The field of education is facing many technical challenges due to the rapid growth of artificial intelligence (AI) work and other technology-based tasks. Educators must be able to not only teach students how to use these new tools, but also need to understand how AI works and how its capabilities are growing.

As AI becomes increasingly sophisticated, educators must keep up with the rapidly changing field in order to equip their students with the skills they need for a successful future. In addition, understanding the ethical considerations involved in using AI is an important part of preparing students for their careers. With this in mind, educators must ensure that they remain knowledgeable about both the technical aspects and moral implications of utilising this new technology.

Educators also face challenges when it comes to finding ways to apply AI knowledge in day-to-day educational practices. As educators strive to stay ahead of the curve, there are some technical challenges that need to be addressed. For teachers and administrators, being able to effectively use AI knowledge in daily teaching practices can prove to be a difficult task.

In order for AI technology to be effectively utilised in education, individuals must first understand how it works and how it can benefit their students. This requires a good understanding of coding, algorithms and data analysis – all skills which many educators may not possess or have access to. Even if they do have the necessary training, they will still need access to expensive resources such as software programs and hardware devices in order for the technology to work properly. This can present an additional financial challenge for schools who may not have the funds required for such investments.

Regulatory Considerations

Companies utilising AI must consider the ethical implications of their technology and how its implementation might affect various stakeholders. For instance, businesses must assess how their AI will interact with other technologies, as well as understand potential privacy or security issues that may arise from using AI-based systems. Additionally, companies should ensure their artificial intelligence models are compliant with any applicable regulations or laws. As such, organisations should continually review existing regulations and be mindful of any new developments that could potentially affect their bottom line.

The use of artificial intelligence offers great potential for businesses seeking a competitive advantage; however, this opportunity does come with certain challenges and regulatory considerations. The implementation process should start by identifying the relevant regulations in your jurisdiction or market sector and then planning how these will be accommodated in the design and development of AI solutions. To ensure compliance, organisations should look at incorporating a risk-based approach when designing an AI system, this means assessing not only the risks posed by the technology but also those posed by its applications. Furthermore, organisations need to ensure that their data is properly managed so as to protect personal information from unauthorised access or misuse.

When it comes to implementing regulatory considerations for artificial intelligence (AI), there are a number of challenges and opportunities that need to be taken into account. Here is how companies should approach this task:

- Companies should first review their existing policies and procedures, making sure they are equipped to handle the complexities associated with AI.

- It is also important to stay informed about legal developments in the field, so that any changes or updates can be implemented as soon as possible.

- Companies should also consider potential risks associated with AI and develop strategies to mitigate them, including privacy protection measures and data security protocols.

- Companies must also assess the ethical implications of using AI, such as potential bias in decision-making algorithms or unintended consequences of certain applications.

Given the potential implications of deploying autonomous systems, governments and regulatory authorities are taking steps to ensure that such systems are used appropriately and ethically. From privacy considerations to potential safety risks, there are numerous issues that need to be addressed in order to maximise the benefits derived from AI technology while minimising any potential harms or unintended consequences.

For these reasons, it is important for organisations considering the implementation of AI-driven solutions within their operations to be aware of the various regulatory considerations involved. This includes understanding applicable laws and regulations as well as obtaining necessary approvals prior to deployment in order to efficiently and safely deploy AI-based solutions. Some nations have taken a more proactive stance, introducing sweeping laws and regulations designed to shape how AI is used within their borders. These rules may address everything from privacy protection to data management, or provide guidance on how government agencies can use AI technology responsibly. On the other hand, other countries may opt for a more hands-off approach, creating guidelines instead of hard-and-fast rules which allow organisations greater flexibility when implementing AI solutions.

Regulatory considerations play a vital role in the international business community. Companies must be aware of the laws, regulations and procedures when operating in different countries. To ensure compliance with regulatory requirements, businesses should consult an experienced international legal advisor. Additionally, companies should remain informed on developments related to current regulations, as well as any new regulations that may come into effect. Regulatory considerations are an important part of any company's operations and should not be overlooked.

Security Concerns

AI has also revolutionised the way businesses are run, allowing companies to remain competitive in an increasingly automated world. However, as with any technology advancement comes security concerns that must be addressed.

AI presents both challenges and opportunities when it comes to security. On one hand, AI is capable of rapidly analysing large datasets and identifying patterns that would otherwise go unnoticed by humans. This can be used to detect potential threats before they become a real problem. On the other hand, malicious actors may use AI-powered technologies to break into systems or launch cyber attacks which can have devastating consequences for organisations and individuals alike.

It is important for businesses and individuals to take precautions when it comes to protecting their data from malicious actors leveraging AI technologies.

The possibilities of AI offer great opportunities for businesses to improve their operations in terms of efficiency and accuracy. However, due to its processing power, it can also pose security risks if not handled properly. With large datasets being analysed by an AI system, there is an increased risk of unauthorised access or manipulation of sensitive information if proper security measures are not implemented. Additionally, the algorithms used in AI systems may have blind spots or errors in identifying patterns which could lead to incorrect decisions being made based on the output data from these systems.

Organisations must be aware of the risks associated with AI-driven systems and take proactive steps to secure their digital assets. This means developing a comprehensive security strategy that addresses the challenges posed by intelligent machines. It is important to consider both technical and non-technical aspects such as user authentication, encryption techniques, system monitoring, auditing procedures and training of personnel. With the right strategies in place, organisations stand to gain significant benefits from leveraging AI while mitigating against potential attacks or data breaches.

Security investments are required to protect data from unauthorised access or use, as well as guard against emerging threats that are increasingly complex due to evolving technologies like AI. Companies should ensure they have suitable mechanisms in place to detect suspicious activities and stop them before they cause any damage. Tools such as encryption, authentication protocols, risk assessment processes, malware prevention solutions must all be utilised for maximum security coverage and protection of confidential information.

While AI presents organisations with an opportunity to stay ahead in the digital world, there are several challenges they must overcome. The following outlines some of the primary security risks posed by AI:

• Data leakage: As AI systems need access to large datasets and sensitive information, there is the potential for data leaks due to poor system design or malicious actors.

• Model poisoning: Attackers can use targeted inputs to trick models into making incorrect decisions, which could have significant implications for any organisation relying on AI-based decision making systems.

• Model theft: Attackers can also steal trained models from an organisation's hardware and then use them as their own or modify them for malicious purposes.

With AI-powered security solutions comes a variety of challenges and opportunities for businesses. Here are some key types of security risks solutions that should be considered:

• Automated Threat Detection – AI can be used to identify malicious activities by learning patterns in user behaviour which can then alert IT teams when suspicious activities occur.

• Real-Time Response – By automatically responding to cyber incidents as they happen, AI can help organisations quickly contain threats and limit damage caused by malicious actors.

- Predictive Analytics – AI can also be used to predict potential attacks based on past events and behaviours, allowing organisations to stay one step ahead of attackers and keep their data safe from harm.

AI security is a rapidly growing field of research and development. It can provide the technology for better detection of malicious cyber-attacks, enable more efficient authentication processes, and use machine learning to identify behaviour patterns that may be indicative of malicious intent. AI security solutions can also help protect online accounts and data as well as detect identity theft. Additionally, AI security systems can help businesses stay compliant with regulations and prepare for potential attacks.

AI in Everyday Life

We are living in a world where artificial intelligence (AI) is rapidly becoming an everyday necessity. From self-driving cars to virtual assistants, AI is transforming the way we interact with technology and making our lives easier than ever before. But what implications does this have for humanity?

What will happen when AI can outsmart humans in almost every field? Will it lead to a utopian or dystopian future? As the use of AI continues to expand, it's important that we keep asking questions about its impact on society and how we can ensure it is used responsibly. With advances in machine learning and natural language processing, AI can be used for both good and bad; but ultimately only time will tell what its true potential will be for humanity.

The potential of AI in everyday life is vast. From facial recognition software that can recognize individuals to virtual assistants like Alexa or Siri that answer questions or complete tasks on command, the possibilities seem endless. Additionally, machine learning algorithms can be used to analyse large data sets quickly and accurately in order to make decisions without human intervention. This could greatly reduce the amount of time it takes to solve complex problems while also cutting down on costs associated with traditional methods.

AI in Everyday Life

The household, a space where we feel the most comfortable and secure in our everyday lives. But what happens when technology invades this personal space and how will it change our lives? With AI making its way into the home, it is now possible to automate mundane tasks and free up time for more meaningful activities.

The implications of having AI in the home are far-reaching. This new age of automation will drastically reshape how we live, work, play and think about our future. We could be walking on a path that leads to a world where machines can do almost anything for us. How will this affect our sense of identity? Will robots take over interactions that used to be performed by people? Where do humans fit into this new equation?

AI-powered devices such as home assistants, autonomous vacuum cleaners and smart appliances are revolutionising the way we interact with technology in our homes. However, these advancements come with their own set of problems that many households will need to address.

Firstly, there can be difficulties when setting up AI-enabled devices. Beyond basic technical issues like connecting to the internet or syncing with other devices, people might have difficulty understanding how to use the device's features properly. This can lead to frustration for those who don't have an aptitude for technology or lack experience interacting with AI devices.

In addition, some aspects of using these devices may make them more difficult than traditional alternatives.

Household security is becoming more and more sophisticated as Artificial Intelligence (AI) becomes an ever-present part of everyday life. With AI being used to monitor home security systems, it's possible for households to become safer than ever before. AI has the potential to track patterns, detect anomalies in behaviour, and even predict risks in order to prevent them from occurring.

This form of monitored home security allows homeowners the peace of mind that their families and belongings are safe from intruders or other threats. AI can be programmed to recognize people who do not belong in the house, identify any out-of-the-ordinary activity, and sound an alert if something seems amiss. Additionally, with features such as facial recognition technology integrated into some systems, protection against unauthorised access is greatly enhanced.

How is it improving households?

AI in Everyday Life is transforming the way we experience our everyday household tasks. From controlling your home's temperature to ordering groceries, AI is revolutionising how we interact with our homes and the items within it. Here are some of the ways AI is making household life easier:

- Smart appliances: Appliances like refrigerators, washing machines, ovens and more are becoming smarter by adding voice control and remote access features. This allows users to check their food supply remotely or adjust temperatures while away from home.

- Voice assistants: Voice assistants can help you do anything from setting reminders to playing music or checking the news, all through simple voice commands. Alexa, Siri and Google Assistant are all great options for quickly getting answers or completing tasks without lifting a finger.

AI in Smartphones

Smartphones, the go-to device for many of us in our everyday lives, have recently been enhanced with the integration of artificial intelligence (AI). AI is a revolutionary technology that has made its way into various aspects of our lives. From virtual assistant applications such as Siri and Alexa to advanced facial recognition features, AI has become an integral part of our daily lives.

AI on smartphones adds a whole new layer of convenience and adaptability to how we use this crucial device. Smartphones can now recognize user behaviours, provide personalised recommendations based on daily habits and preferences, as well as accurately anticipate what users want or need before they even ask for it. All these capabilities are thanks to the increasingly advanced capabilities of AI technologies embedded into today's smartphones.

These AI-based software can help make life easier by understanding user input and providing more intuitive solutions.

Smartphone virtual assistant applications use machine learning algorithms to interpret voice commands, identify patterns in user behaviour, and provide contextual responses that anticipate the user's needs. For instance, they can detect patterns in search queries or conversations to offer more relevant information or recommendations. Additionally, AI-powered assistants can be customised based on individual preferences, allowing users to tailor their experience for greater convenience.

AI technology is quickly evolving and has already been integrated into many aspects of our lives. At the heart of these advanced applications is machine learning (ML) technology, which enables virtual assistants to understand natural human language and respond appropriately. ML algorithms can process huge amounts of data quickly, allowing virtual assistants to make informed decisions without human intervention. With this advanced level of automation at their disposal, users can easily access information or execute specific tasks with minimal effort.

Machine learning algorithms are a type of AI that enable computers to learn and improve on their own without relying on explicit instructions from humans. In this article, we will explore the various machine learning algorithms being used to power applications across industries.

Machine learning algorithms can be divided into three broad categories - supervised, unsupervised and reinforcement learning. Supervised learning involves training a model with labelled data sets, where the output is known in advance; unsupervised learning relies on unlabeled data; while reinforcement learning uses rewards and punishments to teach AI systems how to behave in certain scenarios. Popular supervised algorithm examples include Support Vector Machines (SVM), Naive Bayes Classifiers, Decision Trees and Logistic Regression.

Unsupervised machine learning algorithms are an invaluable tool for developing virtual assistants. With the widespread use of AI in everyday life, these algorithms have become increasingly important for providing users with tailored experiences. Unsupervised learning is a process that allows AI systems to learn without any prior data or labels, enabling them to discover hidden structures and features within large datasets. This type of algorithm is used in virtual assistants to analyse user interactions, context and preferences so they can respond appropriately to requests. By leveraging natural language processing (NLP) techniques and deep neural networks (DNNs), these algorithms can understand subtle nuances in user input that may otherwise be overlooked by humans. Additionally, unsupervised learning helps virtual assistants recognize patterns within customer data as it continues to accumulate over time, helping them provide better customer engagement for improved customer satisfaction levels.

The future of smartphones is an exciting prospect, with many possibilities. Here are 8 ways in which AI could dramatically impact everyday life:

- Smartphones will become more intuitive, able to predict and respond to user needs without explicitly asking for instructions.

- Predictive analytics will anticipate the user's needs and provide information directly on the device – from news updates to personalised product suggestions.

• Automated assistance systems that can handle a range of tasks such as scheduling appointments or responding to messages will be available on most devices.

• Natural language processing enabled by AI-driven algorithms will allow users to communicate with their phones via voice commands instead of typing out instructions manually.

Automation and Robotics

Automation and robotics are two key areas where AI is making a big impact on how we live, work and play. From self-driving cars to robotic vacuum cleaners, AI technology has the potential to revolutionise many aspects of modern society. It can also free up humans for more creative pursuits by taking over mundane tasks such as sorting mail or stocking shelves.

At the same time, automation and robotics will bring about new challenges that need to be addressed. These range from ethical concerns such as job displacement to technical issues related to accuracy and reliability in automated systems. On top of this, there are questions around who should be responsible for any mistakes made by robots or AI-based algorithms.

Robotics is a field of study that has been developing since the early 20th century. It involves the use of artificial intelligence (AI) to control technology and automate processes for everyday life. AI has come a long way in robotics, allowing for machines to complete more complex tasks than ever before.

From industrial robots used to manufacture products on assembly lines, to self-driving cars and personal assistant devices like Amazon's Alexa, AI is rapidly transforming our everyday lives. Robotics experts are now able to create autonomous machines that can interact with their environment and move from one place to another without any human input or interaction. These robots are designed with sensors and algorithms that allow them to identify objects in their environment, respond accordingly, and take decisions based on predefined criteria. By using machine learning techniques, they can even recognize faces or other patterns of movement in order to better understand their surroundings.

By leveraging artificial intelligence (AI) and machine learning algorithms, robots can be taught to perform complex tasks with greater accuracy than ever before. This technology can be used to automate mundane jobs such as sorting items, controlling machines, and organising materials. Additionally, AI-enabled robots can aid humans in dangerous environments such as hazardous waste management or deep-sea exploration where human presence is not advisable.

Robotics technologies are also being used to enhance the performance of autonomous vehicles. By utilising machine learning algorithms, these vehicles can detect traffic signals with improved accuracy and identify obstacles at much greater distances than conventional systems. Moreover, they are able to make decisions faster by using powerful AI software that processes data from multiple sensors in real time.

I-assisted human-robot interaction (IHR) is an emerging technology that is becoming increasingly more important in the field of artificial intelligence (AI). It combines both robotics and AI to enable machines to interact with humans in a natural, intuitive way. By leveraging AI algorithms, robots can be equipped with sophisticated decision-making capabilities and understand complex tasks from spoken commands or visual cues.

The potential applications for IHR are vast and far-reaching, ranging from home automation to healthcare assistants. In everyday life, IHR could be used for tasks such as helping seniors with grocery shopping or providing support for people living with disabilities. As technology continues to advance in this space, the possibilities for what AI can do will expand even further, making it an integral part of our daily lives.

AI in Healthcare

Artificial Intelligence (AI) in Healthcare is an emergent field of study that explores the application of AI technologies to improve the efficiency and efficacy of healthcare delivery. AI-enabled healthcare solutions seek to leverage data, algorithms, and machine learning to automate administrative tasks, facilitate clinical decision support and enable predictive analytics for population health management. AI's potential in healthcare is virtually limitless due to its ability to process vast amounts of data quickly and accurately. As such, AI technology has the potential to revolutionise the way health services are provided, while also reducing costs associated with healthcare delivery.

AI can also be used to analyse large datasets quickly and accurately in order to identify patterns that could help identify diseases earlier or predict treatment outcomes more accurately than traditional methods. Furthermore, AI algorithms can help clinicians make more reliable and informed decisions regarding diagnosis and treatment plans by analysing patients' profiles based on age, gender or other relevant criteria.

AI-enabled healthcare solutions are a novel approach to ameliorating the traditional methods of providing medical care. As such, they seek to leverage the advantages of data and algorithms in order to optimise the efficacy of patient care.

Such solutions are predicated on the notion that algorithmic processing of medical data can be utilised to generate better health outcomes by allowing for a more detailed analysis of individual patients and their respective needs.

AI-driven predictive analytics is a form of data analysis that utilises machine learning algorithms to generate insights from existing datasets. By leveraging AI capabilities, such as natural language processing (NLP), computer vision (CV), and deep learning techniques, the goal of predictive analytics is to accurately forecast future outcomes based on a set of predetermined variables. Furthermore, AI-driven predictive analytics allows for the identification of previously undiscovered patterns in large datasets that may have gone unnoticed by traditional methods.

AI can be utilised to help healthcare professionals in a variety of ways, from predictive analytics to natural language processing. AI-based algorithms are able to identify patterns in medical data that could not be identified by humans and could potentially be used for early diagnosis and disease monitoring. Additionally, AI can be employed to generate insights from large datasets, such as electronic health records, providing physicians with more comprehensive patient information.

One major drawback of using AI in healthcare relates to accuracy and reliability. Many machine learning algorithms do not consider external factors such as environment, lifestyle or medical history when making decisions on diagnoses or treatments which can lead to inaccurate results.

Additionally, AI systems may produce different outcomes depending on training data used which can result in inconsistent performance if not regularly tested and updated by a trained medical professional.

Finally, there are legal implications associated with using AI in healthcare which must be considered before implementation. The concern is the potential risk of infringing on patients' privacy rights by using data collected through AI systems. Data gathered through AI may be used for purposes outside medical treatment, such as marketing or research without the patient's explicit consent which could lead to potential litigation. Additionally, there is a risk that physicians may become too reliant on automated decisions rather than independent assessments which may result in lower quality care and potential liability in cases where incorrect diagnosis or treatments are given due to errors within the AI system itself.

The integration of AI into everyday life is already, and will continue to, revolutionise the way humans interact with technology. As machine learning algorithms become more powerful and versatile, they will have an even greater role in our daily lives. Already, AI has enabled us to communicate faster and more efficiently than ever before.

Moreover, AI technologies are being increasingly used for medical research and diagnostics, allowing for earlier detection of diseases and more accurate diagnoses.

Business Implications

With the rise of artificial intelligence, a new wave of business implications have emerged. Artificial intelligence or AI has been increasingly incorporated into various areas of the business world, such as customer service, marketing and sales, and operations. AI is impacting businesses in terms of streamlining processes, improving customer satisfaction and driving innovation. By understanding the potential implications that artificial intelligence can have on a business's bottom line, organisations can better prepare for changes to come.

AI is becoming more available to businesses at an affordable cost which makes it accessible to even small companies. It helps automate certain manual tasks that were previously handled by employees thus reducing operational costs. With AI-driven insights provided from data analysis, businesses are able to make better decisions based on accurate information about their customers and markets which enhances productivity and efficiency.

Impact on Operations

AI is proving to be a strong tool for enhancing operational efficiency and productivity. It is expected to have significant implications for business operations going forward.

One of the most notable impacts of AI is its ability to automate processes and handle routine tasks with precision and accuracy, freeing up time for other activities. This could lead to increased efficiency in some operations or the development of new ones through the use of automated systems. Additionally, AI can also be used in data analytics, providing insights into customer behaviour that would otherwise remain hidden from view. This could help businesses make more informed decisions as they consider their future strategies and plans.

Moreover, AI can provide real-time feedback on operational performance which can enable organisations to constantly monitor their practices and optimise them accordingly.

It can help minimise operational costs by automating processes and performing tasks more efficiently than human employees. For instance, AI can be used to pre-screen job applicants, allowing employers to save on time and money spent on interviews and other recruitment activities. Additionally, with its ability to sort data faster and more accurately than humans, AI can be used by companies to identify trends that would have otherwise gone unnoticed. This could potentially lead to cost savings due to increased efficiency as well as better decision making based on consumer insights.

Incorporating it into operations security has had a major impact on businesses, both large and small. AI technology can detect potential security threats quickly and accurately, thereby reducing the potential for financial loss or data breaches. It can also help to automate processes and reduce manual labour costs associated with security protocols. However, it is important to recognize that such systems require carefully planned implementation in order to ensure their effectiveness in a business environment.

When incorporating AI into operations security, organisations must consider both technical and procedural aspects of the system in order to ensure its success. On the technical side, AI requires an extensive computing infrastructure that allows it to process vast amounts of data quickly and accurately identify any potential risks or anomalies.

Improving operations within a business can be difficult, especially when considering the implications of how to best utilise resources. However, an innovative approach that many businesses are taking is utilising artificial intelligence (AI) to help streamline their processes and increase efficiency. AI algorithms are useful for analysing data in order to optimise operations, while giving businesses the opportunity to make more informed decisions that can have positive business implications.

● Understand the business implications of operations: Many businesses are increasingly dependent on operations for success, so it's important to fully understand the implications of operations in order to make improvements.

● Analyse current processes and procedures: By studying what processes and procedures currently exist, you can identify potential areas of improvement that may lead to better operations.

● Leverage technology: Technology such as artificial intelligence (AI) can be used to streamline existing processes and automate tedious tasks, leading to improved efficiency and cost savings.

● Develop a plan: Make sure you know exactly what steps you need to take in order to improve your business operations before implementing any changes or upgrades.

Impact on Workforce

Companies are now utilising AI capabilities in order to automate many processes, including customer service and data entry. This can lead to reduced costs for businesses by eliminating the need for manual labour or human resources.

At the same time, the introduction of AI has led to a shift in the skills required from workers. As automation becomes more prevalent, employers are looking for employees who have strong analytical skills and are able to use software programs effectively. Additionally, employers have begun placing greater emphasis on creativity and problem solving abilities rather than technical knowledge when recruiting new staff members. Thus, there is an increasing demand for professionals who possess these qualities in order to remain competitive in today's job market.

The most notable change with the rise of AI is its potential to create jobs through automation, however it can also lead to job displacement for certain sectors and professions. The fear of job displacement has triggered a response from businesses to develop strategies that embrace digital transformation while still considering the impact on their employees. As organisations strive to stay ahead they must assess which areas can benefit most from AI-driven solutions, while ensuring current workers are trained up in new skill sets required by this technology.

Job loss due to the implementation of artificial intelligence (AI) has become a growing concern among workers across many industries. While AI-driven automation can improve efficiency and productivity for businesses, it also raises questions about its implications on jobs and the economy.

For businesses, AI may provide an opportunity to increase profits by reducing labour costs through automation. Automation can be used in a variety of ways from streamlining manufacturing processes to replacing customer service representatives with chatbots that offer immediate responses to customer inquiries. In addition, AI technology could potentially help businesses enhance their operations by collecting data more quickly and accurately than humans are able to do so.

However, there is no denying that job loss is one of the major potential drawbacks associated with the use of AI technology in business environments.

Businesses must take proactive steps to prevent job loss in order to protect their bottom line, staff morale, and reputation. To do this, businesses should first accept that eliminating jobs due to automation is inevitable. It's important to have a clear plan in place so that employees can be retrained or transitioned into new roles when needed.

When considering AI-based solutions, businesses should also focus on creating jobs rather than replacing them. Companies should consider how AI could help increase efficiency while still allowing workers to remain employed and productive. They may need to invest in retraining programs or create new positions such as software engineers or data analysts specifically related to the use of AI technology within the business.

Research suggests that job happiness can have a significant impact on employee productivity and overall business success.

Employees who are satisfied with their jobs tend to be more productive and engaged, which can lead to improved business performance. Additionally, when employees feel fulfilled by their positions, they may be more likely to stay with the company longer-term. This has implications for businesses in terms of recruiting costs and turnover rates. AI can help employers gain better insights into employee satisfaction levels as well as trends in job engagement over time, so they can better respond to employee needs and create a positive work environment.

Impact on Marketing & Advertising

The effects of artificial intelligence on advertising and marketing methods are so profound that print, radio, and television advertisements now appear to be relics of the past. Industry experts are working in house, particularly those that are investigating AI, to develop better forms of advertising that leverage this technology. While predictive analytics have already been used in commercial applications, the process of personalising campaigns is new.

This has enabled marketers to analyse data more quickly and accurately than ever before, from customer surveys to product reviews, providing invaluable insights into consumer behaviour. Moreover, AI-driven programs can now generate personalised ad campaigns that target specific demographics more precisely, thereby leading to increased ROI for businesses. Additionally, AI automation also allows companies to streamline their marketing processes by automating mundane tasks such as social media scheduling and email campaigns.

Overall, the advent of AI offers businesses significant opportunities in terms of optimising their marketing strategies and expanding their reach with targeted ads.

Making the most of marketing strategies is key for businesses to gain a competitive edge and increase their profits.

Optimising these strategies is essential to ensure success and maintain longevity in the industry. Businesses need to consider the implications of their marketing decisions before executing them in order to optimise their strategies and achieve desired outcomes. Artificial intelligence can be used as an effective tool for businesses seeking to leverage information and insights on customer behaviour, preferences and buying trends in order to optimise their marketing strategies.

AI is able to quickly analyse customer data, helping companies target their ideal audience more accurately. This technology can provide valuable insights into consumer behaviour as well as help businesses optimise their campaigns for maximum ROI. As a result, AI-driven advertising is expected to fuel a massive growth in the marketing industry over the coming years.

Businesses that invest in AI-based advertising technologies can benefit from powerful data analytics capabilities, allowing them to better understand consumer trends and preferences. Companies are also leveraging AI to create personalised ads tailored to individual customers' needs. With these insights in hand, marketers are able to craft highly engaging ad messages that drive conversions and increase sales for their clients or brands.

AI has enabled companies to create personalised advertisements tailored to each customer's individual needs. This trend, which began with the introduction of targeted ads based on browsing history and search engine queries, is now becoming even more sophisticated.

AI-driven models are able to track customer behaviour in real-time, allowing businesses to deliver a highly customised experience that meets their customers' exact requirements. As a result, companies can create marketing campaigns that are much more effective at generating leads and sales. Additionally, AI-driven personalisation helps businesses save time and resources as they don't need to manually analyse large amounts of data.

Businesses today are consistently looking for ways to optimise processes and cut costs. One way organisations can do this is by leveraging the power of artificial intelligence (AI) to automate content production. AI-powered automation offers a number of advantages, including cost savings and improved efficiency.

By automating mundane content production tasks such as asset creation, editing, curation and publishing, businesses can save time and resources while ensuring accuracy and consistency in their output. The use of AI also helps reduce errors that might otherwise slow down the process or lead to costly mistakes. Furthermore, AI-driven automation allows businesses to scale quickly with minimal effort – freeing up valuable staff time for more high-value activities such as strategic planning or customer engagement initiatives.

Benefits of AI Adoption

AI adoption provides businesses with a wide range of advantages, from cost savings to enhanced customer service. From streamlining processes and expanding market reach, to improving decision-making capabilities and forecasting potential outcomes, Artificial Intelligence (AI) technology can bring real business benefits.

The introduction of AI solutions can significantly reduce the amount of manual labour required for many operations, allowing businesses to save costs on labour while increasing efficiency through automation. Furthermore, AI solutions can boost customer experience by enabling businesses to offer more accurate and personalised responses to customers' queries. By leveraging AI-driven analytics tools, companies can acquire valuable insights into customer behaviour and preferences which they can use to tailor services according to individual needs and preferences. Moreover, AI solutions help companies make data-driven decisions quickly by predicting future trends based on past performance, resulting in improved decision making capabilities that lead to better results over time.

The use of AI in the customer experience space has several key business implications. For example, AI can be used to personalise the customer journey by leveraging data to recommend products or services tailored to a particular customer's preferences and needs. Additionally, AI-driven automation can significantly reduce operational costs while increasing efficiency throughout the organisation. Finally, through powerful analytics capabilities, AI can provide real-time insights into how customers interact with company offerings so businesses can quickly implement changes or improvements as needed.

All in all, incorporating AI into their customer experience strategies could prove beneficial for businesses looking to stay ahead of their competition and create a more satisfied base of customers.

The algorithms are capable of processing large amounts of complex data to identify patterns and relationships between various factors that cannot be easily identified by humans. This allows AI systems to make accurate predictions about the future based on past performance. Businesses can use this technology to forecast how their industry will evolve, adjust their strategies accordingly, and ultimately remain competitive in an ever-changing marketplace.

This type of predictive analysis powered by AI also enables organisations to quickly respond to market changes and capitalise on emerging opportunities without sacrificing long-term growth goals or overall objectives.

Market changes:

- Businesses are rapidly developing strategies to respond quickly to market changes in order to stay competitive.

- To do this, they are turning to the use of artificial intelligence (AI) to gather insights and analyse data faster than ever before.

- AI enables businesses to identify how customer preferences and industry trends are changing, as well as how their competitors are responding.

- With AI-powered analysis, businesses can quickly detect changes in the market and determine the best course of action for their organisation.

- By utilising AI tools such as machine learning algorithms, businesses can more accurately evaluate customer sentiment and predict future trends with greater accuracy than ever before.

- This allows them to make decisions quickly based on data-driven insights instead of relying solely on intuition or guesswork when responding to market changes.

Challenges to Adopting AI

Businesses are interested in learning more about the cost associated with implementing AI technology. The costs of training employees on using the technology and purchasing AI infrastructure along with hardware or software are significant. Furthermore, managers should prepare for the potential outcomes associated with hiring AI specialists.

Businesses around the world are beginning to realise the potential of artificial intelligence (AI) and its implications for their operations. AI has the potential to revolutionise how businesses operate, from streamlining processes to improving customer relations. However, with any new technology comes questions about cost implications. Business leaders must understand what they will pay for implementation and whether the benefits outweigh the associated costs before investing in AI technology.

AI implementation costs can take many forms such as developing algorithms, hiring specialised staff or purchasing software licences and hardware. Additionally, businesses must consider ongoing maintenance costs such as licensing updates or upgrades for existing software and hardware as well as regular training for staff on new technologies that have been implemented. It is important for businesses to fully understand all potential cost implications prior to making an investment in

AI technology so that expectations can be managed effectively and any ROI can be monitored over time.

The costs associated with implementing AI software can vary greatly depending on the scale of implementation. This may include upfront costs such as hardware purchases or infrastructure setup, as well as ongoing investment for maintenance and additional features over time. Businesses must weigh the benefits that come from using AI against these financial investments to determine if it fits within their budget and provides an appropriate return on investment.

A huge advantage to using AI is its ability to automate mundane tasks that would otherwise take up significant human resources.

The cost of AI software depends on several factors, such as complexity level and data storage requirements. For example, a basic AI model may require minimal upfront cost in terms of hardware or computing resources but could still require considerable investments in terms of manual labour and training hours. On the other hand, more complex models may require significant investments in both hardware and personnel for proper usage. Additionally, ongoing maintenance fees must also be factored into any budget considerations when evaluating AI software costs.

AI technology has already revolutionised the way businesses operate and handle their data. As AI technology continues to evolve, more and more companies will adopt it as part of their daily operations. We are only at the beginning of a new era where AI technology opens up new possibilities for business and unleashes unprecedented opportunities for growth. Companies must stay informed on the latest AI advancements to ensure that they are taking full advantage of its capabilities.

Impact on Education

The advent of Artificial Intelligence (AI) has presented a paradigm shift in the educational landscape, revolutionising the way that students learn and teachers teach. AI-driven technologies such as machine learning algorithms, natural language processing and deep learning networks have been applied to various aspects of the educational experience including assessment, data analysis and personalization. The use of AI in education has enabled more efficient and effective instruction, allowing for greater student engagement and improved outcomes.

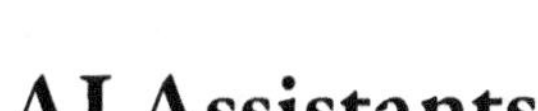

AI Assistants

Artificial Intelligence (AI) is revolutionising the way we interact with technology, and this is especially true when it comes to AI assistants. In terms of education, these AI assistants are set to become an increasingly important tool for students, teachers, and administrators alike.

AI assistants provide a personalised digital learning experience that can be tailored specifically to the needs of each student. With their help, teachers have access to greater amounts of data on individual learners which allows them to assess progress more accurately and provide better instruction in real-time. Furthermore, AI technology can be used in educational settings as a powerful aid for automating administrative tasks such as grading papers or organising online classes.

In short, AI has immense potential when it comes to improving education outcomes both at the classroom level and beyond.

One example of an education AI Assistant is Google Classroom's virtual assistant 'G Suite for Education'. This assistant uses machine learning algorithms to understand student questions and can provide suggested answers based on those queries. It can also be used by teachers to create online courses and lesson plans. Additionally, it has a feature that allows students to submit their work digitally directly from the assistant which helps reduce paper waste in classrooms.

The AI Assistant provides assistance via natural language processing, meaning that users can simply ask questions and receive answers without utilising complex commands or codes. The assistant will search through Google Drive for documents and other online resources, as well as offer direct guidance on the steps necessary for completing tasks or projects. It also has the ability to set reminders for assignments, upcoming events and tests so that students never forget important deadlines.

For teachers, this AI Assistant offers assistance with managing student progress by providing insights on individual performance within a class.

Automation and Grading

Automation and grading are two concepts that have become increasingly important in the modern education system. As technology advances, so does our ability to leverage artificial intelligence (AI) for use in classrooms. By automating tedious tasks such as grading, teachers can focus more on teaching and less on paperwork, giving students more time to learn.

The use of automation and AI offers numerous benefits when it comes to assessment and grading. For instance, automated systems allow instructors to grade essays quickly while providing detailed feedback. Automated systems can also be used to generate personalised assessments tailored specifically for each student's learning style and needs. Additionally, many schools have adopted interactive feedback tools that allow teachers to monitor individual student progress over time with data-driven insights.

In modern education, schools have adopted interactive feedback tools to help students improve their learning and overall performance.

The use of these artificial intelligence (AI) enabled tools helps in providing personalised feedback for each student, relying on data collected from the students' interactions with digital content. This technology is designed to analyse and identify areas where a student may be struggling or excelling, as well as locate any knowledge gaps they may need to fill.

The feedback provided can range from simple hints and tips on how to improve their understanding of a concept, right through to more detailed and personalised recommendations on how best to progress forward.

Furthermore, these AI-powered tools are designed not just for giving feedback, but also for tracking a student's progress over time. This allows teachers to better identify potential issues before they become too large or unmanageable.

Grading

Grading is a vital part of the education system, and its importance has only increased in recent years. With the introduction of artificial intelligence (AI) into the grading process, teachers are able to grade more effectively and efficiently. AI-based grading systems can identify patterns in student work that would be difficult for humans to spot, as well as providing data-driven feedback in real time.

Using AI to grade students' work has advantages beyond improved accuracy and efficiency. By automating certain aspects of grading, teachers have more time to focus on other important tasks such as meeting with students or preparing lessons. Additionally, AI-based systems can provide personalised feedback tailored specifically to each individual student's needs, helping them improve their understanding of course material faster than ever before.

AI Tutors

The world of education is rapidly evolving, and artificial intelligence (AI) has opened up a new realm of possibilities. AI tutors are now emerging as powerful tools to help students learn faster and more effectively.

These digital tutors have the potential to revolutionise the way we learn by providing personalised instruction tailored to each student's individual abilities and needs. They can identify learning gaps, track progress, and provide real-time feedback on performance. This technology can also be used in conjunction with traditional teaching methods, helping teachers deliver more effective lessons while freeing up their time for other tasks.

At the same time, AI tutors come with their own set of challenges such as privacy concerns or potential bias in algorithms that could negatively impact learning outcomes.

AI tutors rely on algorithms to create their lesson plans and analyse student performance data. These algorithms could potentially introduce bias into the system due to incorrect assumptions made by the programmers who created them. This could lead to skewed results or misunderstandings about student progress which might have a negative effect on their learning performance.

Is this really benefiting students' learning performance?

The intention of AI tutors is to provide personalised instruction that can help students learn material more effectively. While they offer an interactive and engaging approach to learning, AI tutors also have some drawbacks. There are concerns that these digital tools may replace human-to-human interaction between teachers and students in the classroom, thus impeding their education development potential. Additionally, there is evidence suggesting that too much reliance on AI tutors may actually lead to decreased knowledge retention for certain topics due to lack of engagement from the student's part.

Some concerns have been raised over AI tutors potentially being used to cheat. AI tutors provide an automated, personalised way for students to learn, but can also be used by unscrupulous individuals to get around academic integrity rules.

Unlike traditional cheating techniques, AI tutors can be difficult to detect because they are designed to mimic human behaviour. These programs can generate answers quickly and accurately enough that a student could use them without risking detection. They also allow users to take advantage of loopholes in the educational system. For example, they could be used to answer test questions or complete assignments without ever having seen the material beforehand. In addition, these programs can be difficult for instructors and administrators to control since it's hard to track user activity on computers outside their direct supervision.

Virtual Reality Experiences

Virtual reality (VR) is becoming a popular way to experience unique, immersive environments. It has the potential to revolutionise how we interact with the world around us, including in education and other areas. With its growing popularity, VR is now being used in conjunction with artificial intelligence (AI).

The use of VR coupled with AI offers an unprecedented level of realism and immersion that traditional methods simply cannot provide. For instance, AI can help create highly interactive learning experiences by simulating real-life scenarios and giving users the opportunity to practise decision making skills in a safe environment. Additionally, AI can be used to monitor user progress, provide personalised feedback and even adapt content accordingly for individual students.

By utilising VR with AI, educators now have access to more engaging teaching methods that are designed to improve student learning outcomes.

Whether it's learning a foreign language, exploring outer space, or experiencing another historic period, virtual reality is proving to be an invaluable tool for educators who want their students to gain deeper insights into the subject matter. The combination of VR and AI allows educators to create immersive simulations and provide personalised feedback in real-time. This level of interactivity makes VR a great addition to any classroom curriculum as it helps foster critical thinking skills while engaging learners with its visual appeal.

One type of activity is the use of AI-driven simulations that replicate complex scenarios like medical emergencies or natural disasters within a safe environment. With this type of activity, students can practise decision-making skills in real-time while interacting with realistic characters and objects. This provides an immersive learning experience for students, which can help them better understand how to respond to these types of situations. Not only does this type of virtual reality training provide invaluable experience in a safe environment, but it also allows educators to customise each simulation and track a student's progress as they interact with it. Furthermore, AI-driven simulations allow educators to easily adjust scenarios in order to challenge learners at different levels and tailor the experience accordingly. This type of technology can be used across all educational levels from kindergarten through college and beyond.

VR schools are using the technology to give students a more immersive learning experience than ever before. By combining traditional teaching methods with VR, educators are able to provide students with an engaging way to learn about the world and its many complexities. It enables instructors to tailor lesson plans specifically for each student's interests and abilities, ensuring they get the most out of their educational experience. VR school will equip learners with valuable skills such as problem-solving, critical thinking and collaboration that they can use throughout their lives. It provides a safe environment where students can experiment without the pressure of real life consequences, giving them an opportunity to develop confidence in themselves and their abilities.

All aspects of the school, from lessons to student interactions, take place in a virtual world. The classrooms are tailored specifically to each student's level of knowledge and allow them to learn at their own pace with personalised instruction. VR schools offer an array of interactive activities that help students explore their interests while honing their skills. From virtual field trips to immersive simulations, there is something for everyone. Students can collaborate on projects with classmates from around the world and engage in meaningful conversations about real-world issues while practising critical thinking skills. With this innovative educational model.

The VR school model offers an intriguing blend of physical and digital learning experiences that allow students to access resources from any location while also connecting them with peers from around the world. Through interactive activities such as 3D simulations, augmented reality tours, and virtual field trips, students can explore concepts with greater depth than ever before. Furthermore, AI capabilities provide teachers with powerful analytics tools to measure student progress and customise lesson plans according to individual needs or interests.

Data Collection and Analysis

Data collection and analysis is a key part in the development of artificial intelligence (AI) in the field of education. As high-achieving AI technologies such as computer vision become more advanced, it's vital to know how data is collected and analysed to help alleviate educational constraints. AI systems can interpret large amounts of data with ease, enabling teachers to make thorough analyses of student behaviour.

Schools, teachers, and researchers must use reliable methods when collecting data as it will be used to inform decision-making processes around students' learning experiences. Data can be collected through surveys, interviews with teachers or students, classroom observations or by analysing digital records such as test scores or other academic activities. It is important to collect information that accurately reflects the current state of a student's learning journey. An accurate analysis must be carried out on data gathered so that a constructive result can be achieved from it.

Data collection is the first step in understanding how well students are doing in their education. This involves gathering information about individual student performance such as test scores, attendance records, grades, etc., as well as larger trends in the classroom or school overall. Through analysis of this data, teachers can identify areas for improvement and create targeted interventions to help struggling learners succeed.

The use of AI further enhances data collection and analysis by allowing for greater accuracy and efficiency when gathering large amounts of information from multiple sources. AI can help to identify, analyse, and interpret complex data sets; however, if the raw data fed into algorithms is inaccurate this can lead to flawed results. To ensure that accurate data and results are obtained from AI applications in education, there are a few key steps which should be taken.

The first step in obtaining accurate results from AI-based educational applications is to acquire reliable sources of data. This means verifying the validity of each source used for both internal and external datasets. Once verified sources have been identified then it's important to ensure the datasets themselves have been accurately assembled by double checking for errors or inconsistency within them before any analysis can begin.

Benefits of AI in Education

The modern education system is changing rapidly with the introduction of artificial intelligence (AI). AI has enabled educational institutions to provide personalised learning experiences for students, creating an environment in which they can learn more effectively. The benefits of using AI in education are multifaceted and far-reaching.

AI technology helps to save time by automating mundane tasks like grading tests and tracking attendance. It also enables teachers to monitor student performance more accurately, allowing them to quickly identify any areas where a student may be struggling or excelling. Additionally, AI can be used to create virtual tutors that are capable of providing individualised instruction and support for students who need extra help understanding a concept or topic. Finally, AI allows for adaptive learning models that automatically adjust the content being presented based on the student's current level of knowledge and progress.

Artificial Intelligence (AI) has revolutionised the education sector by introducing innovative tools and technologies that improve the learning process. There are numerous advantages of AI in education, such as:

- Improved Accessibility – AI solutions enable students to access educational content from anywhere, regardless of their geographic location or financial constraints. It can even be used to provide personalised instruction tailored to each student's needs.

- Enhanced Learning Experience: AI can help make learning more engaging by creating interactive experiences that stimulate curiosity and foster knowledge acquisition more effectively than traditional methods.

- Automation: AI automation solutions can streamline administrative tasks, such as grading tests, tracking attendance, and managing digital course materials for teachers and administrators alike. This frees up valuable time for educators to focus on teaching rather than paperwork.

AI has had a significant influence on the field of education. AI-powered technologies have permitted students to access learning resources more easily, while AI-driven assessments and analytics are giving insights into how students best learn. In addition, AI is transforming the way in which teachers teach and providing them with more individualised lesson plans. Ultimately, AI is playing a role in teaching kids by the improved personalised feedback.

Impact on Jobs

Artificial intelligence has significantly impacted the job market, transforming industries while leaving many people wondering about its long-term effects. Modern AI is explanatory of a company's rise or decline, with many workers wondering what the future holds for their jobs.

As artificial intelligence keeps growing more advanced, businesses are experimenting with it to increase productivity. This usually means that certain jobs can now be done by machines instead of people. In some cases, this has caused an increase in automation, whereas in others, it has resulted in workers losing their jobs. On top of that, AI tends to be an ever present concern. The growing impact of artificial intelligence on the job market does not seem likely to reverse anytime soon.

In fact, the jobs that are most at risk of getting taken over by AI are those which involve a lot of repetitive and predictable tasks.

The good news is that for those who fear their job is going to be replaced by a robot, there are some who think AI will actually create more jobs than it takes away.

A report found that 38 percent of all current jobs in the U.S. are at risk of being replaced by AI and other technologies over the next 20 years, but this will be offset. By contrast, an estimated last year that as many as 7.5 million jobs in the U.S. would be lost to automation over a period of 12 years. Roughly 45 percent of the activities people are paid almost $16 trillion a year to do in the global economy have the potential to be automated by adapting currently demonstrated technology.

The economic toll would include the loss of more than $2 trillion in annual wages and 2.5 million jobs, with China and India being hit hardest.

Impact on Low-Skilled Jobs

As the world of technology advances, so does the impact on low-skilled jobs. In particular, artificial intelligence has been shown to have a significant effect on many occupations that rely heavily on manual labour and require less education. This includes many manufacturing jobs, and even some lower-level management roles. The reason for this is that robots and artificial intelligence can complete routine tasks much more quickly than humans. They have the ability to 'learn' as they work and thus get better at a task over time. Ultimately, that means they can complete tasks more quickly and at a lower cost than humans.

Research predicts that AI will most impact low-skill workers, who are already struggling. This could lead to a drastic decrease in job security for those in low-skill positions and could cause an increase in unemployment rates across the country. Low-skilled jobs are not only affected by AI's potential displacement of human workers; they are also impacted by its ability to automate certain tasks such as customer service or delivery operations which can lead to fewer job opportunities available.

As AI becomes more integrated into our daily lives, the need for specialised skills will increase. This means that jobs requiring complex problem-solving, critical thinking, creativity and social intelligence will be among the most difficult to automate.

AI will also give rise to new areas of innovation, as developers struggle to overcome the challenges that arise from having machines with a human-like ability for perception and decision-making.

The impact of artificial intelligence (AI) on low skill jobs is a major concern for many. AI technology has been rapidly gaining traction and is increasingly replacing human workers in many industries, including those requiring low skill labour. This trend has the potential to dramatically reduce employment opportunities in sectors that have traditionally supported individuals without college degrees or specialised skills sets.

Recent studies indicate that as much as 47% of current job roles could be replaced by automation within the next two decades. This could be particularly damaging for low-skill occupations, such as sales and service jobs, which are more easily performed by machines than those which require higher cognitive functions. For example, AI has already taken over customer service tasks such as answering emails and responding to web chats, leaving fewer opportunities for humans to fill these roles.

But there is a silver lining to this cloud – many employers are realising that AI and automation can reduce costs, freeing up resources to invest in higher-skilled positions. For example, it's possible for machines to undertake basic data entry or cleaning tasks more efficiently than humans. In turn, this helps businesses stay competitive by increasing productivity while decreasing labour costs.

As AI advances, more and more jobs become automated, leaving low-skill workers without opportunities to thrive.

But this doesn't have to be the case! Upskilling, the process of teaching employees new skills to keep up with advances in technology, is an effective way of ensuring that these vulnerable workers remain employable. Through upskilling initiatives, lower-skilled workers can gain access to higher paying jobs and economic security. Employers benefit too, upskilling their existing workforce can help them stay competitive in today's rapidly changing job market.

Upskilling will play an increasingly important role as AI continues to disrupt the world of work.

There are a variety of methods available to anyone looking to develop their skills without having a specialist education or experience. For instance, digital media and programming courses can be taken online or through local courses at libraries and community centres. Additionally, volunteer work is an excellent way for individuals to both give back while also enhancing their resume credentials which can open up more mid-level employment opportunities down the line.

Impact on Mid-Level Jobs

The advent of artificial intelligence (AI) has created a buzz in the global economy. As AI technologies become more ubiquitous and sophisticated, there is an increasing potential for their impact on mid-level jobs.

Mid-level jobs are those that require higher skill levels than entry-level positions but do not necessitate advanced technical know-how. They often involve clerical tasks, customer service, and general management duties. With AI's ability to automate these types of tasks, there is a growing concern about the displacement of human labour in the workplace. As a result of this shift in entry-level qualifications, job seekers must be prepared to demonstrate advanced proficiency in programming languages along with applicable software engineering experience. Additionally, new technologies such as machine learning and natural language processing are becoming increasingly important for entry-level positions requiring data analysis or customer service roles that involve providing information quickly and accurately. Companies may also require prospective employees to showcase their ability to think critically when it comes to problem-solving tasks or designing algorithms for certain processes.

The effects of AI are still being researched; however some experts predict that automation could lead to a decrease in job opportunities at the mid-level while creating new roles requiring advanced technical skills such as programming or data analysis.

The jobs that are most likely to be impacted by AI include: Telemarketing and many administrative support roles such as customer service representatives, data entry clerks, secretaries and travel agents. As well as roles in retail, financial services and insurance.

AI is also likely to impact many jobs in the legal profession, and sectors such as healthcare could be affected by improved diagnosis through AI. Some jobs that involve making decisions on behalf of others. such as air traffic control could also be affected.

Mid-level positions often require a certain level of skills that cannot be automated by AI. To ensure success in such roles, it is necessary for individuals to possess a variety of technical and interpersonal abilities.

The most important mid-level qualifications include problem solving, collaboration, multitasking, communication, critical thinking, creativity, and adaptability. Problem solving allows individuals to troubleshoot issues quickly and accurately; collaboration enables employees to work together effectively; multitasking helps them juggle multiple tasks at once; communication facilitates understanding between team members; critical thinking helps them make informed decisions; creativity sparks innovative ideas; and adaptability allows workers to adjust quickly when circumstances change.

Impact on High-Skilled Jobs

The impact of artificial intelligence (AI) on high-skilled jobs is a growing concern among professionals and industry experts alike. AI has the potential to revolutionise businesses and organisations, but it also carries with it the risk of replacing human professionals in certain roles. With advances in technology, AI is playing an increasingly important role in many industries, from finance to healthcare.

For high-skilled workers, this could mean that their job may be at risk if AI algorithms can do their job faster and more accurately than humans. On the other hand, some experts caution against fearing this technology too much as there are still tasks that require a human touch, such as creative thinking or complex problem solving. In addition, AI can help open up new career opportunities for those willing to learn how to use and develop these technologies.

High-skilled jobs are being affected by the increasing use of Artificial Intelligence (AI). Here's how:

- AI is also taking over more complex tasks such as data analysis and medical diagnostics that once required a high level of expertise from humans.

- As more businesses rely on machine learning technology for their daily operations, some lawyers and accountants may be at risk of losing their positions to computers that can process paperwork faster and with fewer errors than humans ever could.

As a result of advances in artificial intelligence (AI), jobs requiring high-skilled labour are increasingly being automated. This shift is having a profound impact on the workforce, as companies from all industries are finding new and innovative ways to use AI technologies to reduce costs, increase efficiency, and optimise processes. As these changes take effect, skilled workers must now adjust their job descriptions and consider additional training in order to remain competitive in the job market.

At the same time, there is an opportunity for those willing to learn how to use and develop AI technologies. With this skill set comes access to new career opportunities which could not have been imagined just a few years ago. Companies are increasingly looking for professionals with expertise in areas such as machine learning, natural language processing (NLP), computer vision, robotics engineering, and more.

High-skilled jobs are becoming increasingly important in the modern economy. With technology evolving at a rapid rate, it is important for individuals to gain the skills necessary to remain competitive in an ever-changing job market.

- Artificial Intelligence: AI is changing the way we work and has become an essential part of many high-tech roles. Understanding basic concepts such as machine learning, natural language processing and computer vision can give you a competitive edge when looking for higher level positions.

● Coding: Knowing how to code is an asset no matter what field you're in, from finance to healthcare. Having coding skills can help you solve problems more efficiently and understand how different systems interact with one another, which could be essential for certain roles.

Benefits of AI for Job Market

AI can potentially provide numerous benefits for employees, employers, and society as a whole.

In order to understand how AI affects job opportunities, it is important to recognize both its potential benefits and risks. On the plus side, AI can automate much of the laborious processes involved in certain types of work such as data entry or customer support. This frees up human employees to focus on more meaningful tasks that offer greater career advancement potentials and higher salaries.

How does it affect small businesses and the job market?

Small businesses are increasingly using resources and job benefits that leverage artificial intelligence (AI). AI offers a range of opportunities for small businesses to streamline their operations, reduce costs and increase efficiency. For employees, AI-powered job benefits make it easier to find jobs, get paid faster and manage workloads more efficiently.

In terms of impact on jobs, AI is revolutionising the way small businesses deliver services and products to their customers. By automating certain tasks such as customer service or data entry, it can free up time for employees to focus on higher value work like product development or strategy initiatives. This shift in focus allows small business owners to better utilise the skills of their workforce while also allowing them to complete tasks much faster than ever before.

How will it impact the job market?

The job market outlook is seen to be impacted significantly by the increased use of artificial intelligence (AI). AI promises to automate tedious and repetitive tasks, reducing the need for human labour in certain areas. This could spell trouble for some sectors that rely on low-skilled labour and require little technical know-how.

At the same time, the availability of AI technology is creating new opportunities in other sectors. For instance, businesses are able to leverage AI in order to gain insights from large data sets and improve their business operations. This presents an opportunity for those with strong technical backgrounds who can develop and maintain AI systems. Furthermore, experts suggest that many jobs will emerge which involve managing AI systems as well as providing services related to machine learning and big data analysis.

Challenges for Job Market

The job market is changing rapidly due to a myriad of factors, including advancements in technology. Artificial intelligence (AI) has drastically impacted the job market and continues to cause shifts in employment opportunities. Specifically, AI is increasingly being used for tasks that were once done by humans, resulting in fewer roles available for individuals seeking work.

In addition to reducing the number of jobs available, AI also impacts which jobs are most sought-after. As machines become increasingly automated and sophisticated, more employers are looking for skills related to technological fluency as well as problem solving and critical thinking abilities. In order to remain competitive in the job market, individuals must possess an understanding of technology or be prepared to quickly learn new skills.

These changes have created a challenging landscape for those searching for employment opportunities; however they have also opened up a range of new possibilities.

AI is being used to automate many tasks that were previously performed by humans. These AI systems require human oversight and training, which means there will be continued demand for skilled workers in the workforce.

The current job market has seen significant changes due to technological advances, with a large impact on roles and opportunities available. Artificial intelligence (AI) has been an important factor in this shift, creating both new jobs and eliminating some existing ones. This is having a profound effect on the job market and the way people are finding work.

One challenge posed by AI is that it can replace many manual labour or low-skill positions that have traditionally provided jobs for certain segments of society. This removes options for those who may not have access to higher education or specialised skills from entering the workforce. It also reduces competition for skilled workers, potentially resulting in stagnation of wages and fewer benefits being offered by employers.

At the same time, however, AI has opened up new opportunities where humans can collaborate with machines to create more complex solutions than either could achieve alone.

The current job market has seen significant changes due to technological advances, with a large impact on roles and opportunities available. Artificial intelligence (AI) has been an important factor in this shift, creating both new jobs and eliminating some existing ones. This is having a profound effect on the job market and the way people are finding work.

One challenge posed by AI is that it can replace many manual labour or low-skill positions that have traditionally provided jobs for certain segments of society. This removes options for those who may not have access to higher education or specialised skills from entering the workforce. It also reduces competition for skilled workers, potentially resulting in stagnation of wages and fewer benefits being offered by employers.

At the same time, however, AI has opened up new opportunities where humans can collaborate with machines to create more complex solutions than either could achieve alone.

Human vs. Machine

The difference between human intelligence and artificial intelligence is a common topic of discussion in the 21st century. It refers to the comparison between human intelligence and artificial intelligence (AI). AI is a company that has been developed to mimic human behaviour, with the ability to think and reason on its own. This technology can be used in many different ways, from simple tasks such as voice recognition to more complex tasks like driving a car.

It all depends on which system will be more productive in the long run. The debate between humans and machines is ongoing because it depends on whether the machines will outperform humans in the long run on complicated tasks. There are some who assert that machines are better in the long run due to their speed and accuracy compared to humans; others believe that humans will always retain their advantage over machines because they have creativity and human qualities that the machines will never have.

AI and Human Interaction

In the current technological landscape, artificial intelligence (AI) is increasingly becoming intertwined with our lives. AI has been used to develop a variety of capabilities ranging from facial recognition software to virtual assistants. As AI becomes more advanced, it raises questions about the future of human-machine interactions and its effect on humanity.

The concept of humans versus machines has long been an area of debate since the beginning of industrialization. While technology can have some advantages such as increased efficiency and productivity, there are concerns about the potential impacts it can have on humans in terms of job security and the potential for automation to replace certain roles that require creativity or critical thinking skills. Additionally, there are ethical considerations surrounding how artificial intelligence could shape decision making processes that could lead to potential discrimination or unequal access to resources and services.

The battle between humans and machines has been going on for quite some time. In recent years, the rapid development of artificial intelligence (AI) has caused a stir in the scientific community, as well as among laypeople. AI is a branch of computer science that studies algorithms and software to create intelligent machines that can act independently from humans.

While many people have expressed concern over the potential competition between humans and machines, others see AI as an opportunity for us to further our knowledge and expand our capabilities in unprecedented ways.

AI can help us with tasks that are too difficult or dangerous for humans to do alone. It assists us in making decisions quickly by detecting patterns and analysing data sets faster than any human could ever do it manually.

AI is a form of computer science that allows machines to operate with intelligent behaviour, enabling them to perform tasks that were previously thought impossible. This technology is being used in a wide variety of sectors, from engineering and manufacturing to healthcare and finance. Despite its many potential benefits, there are still questions about how AI will ultimately affect human societies.

On the one hand, AI can be helpful in addressing challenging problems such as climate change or pandemics by providing more efficient solutions than humans alone could achieve. It can also automate certain processes that make our lives easier; for instance, self-driving cars can help reduce traffic accidents due to their ability to process data faster than a person ever could.

AI has the potential to revolutionise the way humans interact with technology and each other. AI systems can provide us with more accurate predictions, better decision-making, and cost savings in many industries. As AI continues to evolve, it will become more integrated into our everyday lives and will undoubtedly shape the future of human societies. We must prepare ourselves for this changing reality by understanding its implications, developing ethical standards, and investing in research and development.

Benefits of Combining AI & Human

One of the most intriguing advances in modern technology is the combination of artificial intelligence (AI) and human capabilities. This powerful blend has opened up a world of possibility, allowing people to solve problems with far more efficiency and accuracy than ever before. By combining AI and humans, both parties can super-charge their abilities to tackle complex tasks, whether it's analysing customer data or searching for new medical cures. The benefits that come from using both humans and machines together are numerous.

AI can take over many mundane tasks which would otherwise require humans to spend time on manually. For instance, if a task requires sorting through thousands of documents quickly without errors, AI can provide an efficient solution that would save countless hours in comparison to having a human doing it alone. Firstly, it can help augment individual capabilities. Machines are able to process data faster than humans and can potentially aid in decision-making. Secondly, tasks that use a combination of human and machine capabilities can be completed faster and more efficiently. For example, if machines are used for mundane tasks such as sorting data or organising files alongside humans, the work will be done quickly with minimal errors. Thirdly, AI can also help reduce costs as machines eliminate labour expenses while improving accuracy. Finally, by pairing human intelligence and Artificial Intelligence (AI), there is potential to create new innovative ideas that neither could have created on their own.

Human disabilities

AI can help by providing enhanced access to technology for individuals living with physical or cognitive impairments. For example, voice-activated systems allow people with limited mobility to control their devices without having to use their hands or body movements. This can be highly beneficial for those who rely on wheelchairs or have other motor impairments. AI can also provide insight into how humans interact with one another in order to better understand disabilities such as autism and cerebral palsy.

The combination of AI and human can provide many benefits to those living with disabilities:

- Automated assistance can reduce the burden of physical and mental tasks on individuals.

- AI-driven interfaces can provide more comprehensive support for communication needs.

- AI-based mobility solutions can help enhance movement capabilities for those with limited mobility.

- Remote monitoring solutions enabled by AI can ensure that specialised care is provided in a timely manner when needed.

Challenges of Human/Machine Interaction

The concept of human-machine interaction has been around for centuries, yet it remains one of the most confounding challenges of modern technology. Human vs. Machines are a continuous tug-of-war between the efficiency and accuracy of artificial intelligence (AI) and the nuanced judgement, intuition and compassion that humans possess.

In many ways, AI has advanced to be far more reliable than humans in certain tasks such as data collection or processing; however, when it comes to decision making or communication with users, machines may still lack a level of empathy that's essential to building relationships with customers. Companies must carefully consider how they can utilise both human and machine elements to create an optimal customer experience while also using AI's capabilities to streamline operations where possible.

The challenge of human-machine interaction is becoming increasingly pertinent as artificial intelligence (AI) technologies become more pervasive in our lives. As the power of AI continues to increase, it has changed the way we interact with machines and vice versa. The relationship between humans and machines has grown complex due to the increasing presence of AI, which can lead to unforeseen issues for users.

One major challenge posed by human-machine interaction is that these machines may be unpredictable or behave differently than what was originally intended. This can lead to user frustration if expectations are not met or errors occur during operation. Additionally, as AI becomes more sophisticated, it can take on tasks that were previously done exclusively by humans; this could displace certain jobs and create economic uncertainty while also changing how people interact with computers in their daily lives.

AI-driven human machine interaction presents a number of challenges for developers and users alike. Although many of these issues are technological in nature, there are also significant social and legal implications that must be taken into account when using AI to facilitate communication between humans and machines. It is vital that we create frameworks to ensure the highest standards of security and privacy, while also optimising user experience.

ABOUT THE AUTHOR

Lyndee Maharaj is an aspiring philosopher whose goal is to understand σοφία. She has a deep passion for learning. A lifelong learner with an interest in Computers, Physics, Mathematics, Languages, Visual Arts, Baking, and Video Games.

Her hobbies include learning ancient languages, baking, painting, Star Wars, cosplay, computers, and music.

Her passion is lifelong learning be it ideas from different disciplines, people from different teams, or applications from different industries. She has strong technical skills and an academic background in engineering, statistics, and computer science.